AF553811

THERMOCHEMISTRY

By

Dr. Praveen Tyagi

Lecturer

D.P.B.S., P.G. College

Anoopshahr, Bulandshahr

(U.P.)

DISCOVERY PUBLISHING HOUSE

NEW DELHI-110002

First Published-2006

ISBN 81-8356-140-3

Published by

DISCOVERY PUBLISHING HOUSE
4831/24, Ansari Road, Prahlad Street,
Darya Ganj, New Delhi-110002 (India)
Phone: 23279245 • Fax: 91-11-23253475
E-mail:dphtemp@indiatimes.com

Printed at:

Amit Enterprises

Preface

This book has been written for the students of under-graduate and post-graduate level of the various universities in India. A special feature of the book is that the text has been illustrated with a large number of line diagrams and the data presented in the form of numerous tables for reference and comparison. In the preparation of text standard works and review by renowned author have been freely consulted and the reference given chapter wise. At the end of the book will be found useful by those who wish to make a more detailed study of the topics discussed.

We are extremely grateful to our respected Managing Director Shri Tilak Wasan, Discovery Publishing House, for valuable and active cooperation. We humble request our students and chemistry teachers to send us their constructive criticism and suggestions which we shall be using in the publication of the next edition.

Author

Contents

Pages

Preface

1. **Thermochemistry** 1

Introduction, Thermochemical Equation, Conventions Used in Writing Thermochemical Equation, Heat of Reaction or Enthalpy of Reaction, Factors on Which Heat of Reaction or Enthalpy of Reaction Depends, Exothermic and Endothermic Reactions, Standard Enthalpy Changes, Various Types of Enthalpy Changes, Experimental Determination of ΔE and ΔHD, Kirchhoff's Equation, The Maximum Flame Temperature, Explosion Temperature, Bond Enthalpies or Energies, Law of Thermochemistry.

2. **The Specific Heat of Solids & Gases** 62

Specific Heat of Solids, Debye Theory, Blackman's Lattice Theory of Specific Heats, Specific Heat of Gases.

3. **Non-Equilibrium and Classical Thermodynamics** 80

Non-Equilibrium Thermodynamics, Theorem of Wiener-Khintchine, Production of the Entropy, Assumption of Local Equilibrium, Fluxes and Forces, Onsagar Relations and Applications, States of Stationary, Introduction to Classical (Statistical) Thermodynamics, Theorems of Probability in Statistical Thermodynamics, Probabilities Calculation, Statistical Postulates, Ensemble Method, Accessible Microstates, Quantum Statistics, Statistical Mechanics, Statistical Equilibrium, Maxwell-Boltzmann Distribution Law, Maxwell-Boltzmann Distribution in Terms of Temperature, Fermi-Dirac Distribution Law, Maxwell-Boltzmann Distribution and Ideal Gas, Electron

Gas, Distribution Law of Bose-Einstein, Some Definitions, Statistics Applied to Molecules, Macroscopic Systems Characteristic Features, Equilibrium State of the System, Gibbs Ensemble, Some Ensembles, Solar Constant, Stefan's Law (Derivation of Mathematical), Cooling From Stefan's Law (Derivation of Newton's Law), Stefan's Law Verification, Photon Gas, Gibbs Function.

4. Kinetic Molecular Theory of Gases **198**

Postulates of the Kinetic Molecular Theory of Gases, Compressibility Factor, Expansivity and Compressibility, Deviation From Ideal Gas Behaviour, Equations of State for Real Gases, Explanation of Behaviour of Real Gases of the Basis of Van Der Waals Equation, Experimental Determination of Critical Constants, Critical Constants, Conditions for Liquefaction of Gases, Principle of Continuity of States, Van Der Waals Equations and Critical State—Relations Between Van Der Waals Constants and Critical Constants, The Principle of Corresponding States, Liquefaction of Gases, Intermolecular Forces or Van Der Waals Forces.

1

THERMOCHEMISTRY

INTRODUCTION

It is well known that nearly all chemical reactions are accompanied by energy changes. These changes appear, ordinarily, in the form of *evolution or absorption of heat. This branch of chemistry which deals with the study of energy changes accompanying chemical reactions is termed as thermochemistry.*

The energy given out during a chemical change appears in the form of heat. While that which is absorbed may be in the form of thermal, electrical or phóto energy. The amount of energy evolved or absorbed during a chemical change always remains same for the same quantities of reacting substances.

The subject matter of thermochemistry is based on the first law of thermodynamics. The energy changes in chemical reactions are generally due to breaking up of the existing bonds between the atoms and the formation of new bonds. Thus, thermochemistry provides important informations regarding the bond energies.

THERMOCHEMICAL EQUATION

An equation which indicates the evolution or absorption of heat in the reaction or process is called a thermochemical equation. For example, the following equation

$$C + O_2 \rightarrow CO_2 + 393.5 \text{ kJ} \qquad ...(1)$$

reveals that when carbon burns in oxygen to form carbon dioxide, 393*5 kJ of heat are set free per one mole of carbon dioxide produced. Similarly, the equation

$$C + 2S \rightarrow CS_2 - 92.0 \text{ kJ} \qquad ...(2)$$

reveals that one mole of carbon combines with 2 moles of sulphur to form one mole of carbon disulphide with the absorption of 92.0 kJ of heat. Thus, the equations (1) and (2) written above are the thermochemical equations.

CONVENTIONS USED IN WRITING THERMOCHEMICAL EQUATIONS

(1) Physical States of Reactants and Products

The physical states of reactants or products have been found to influence the magnitude of heat evolved or absorbed in the reaction.

Therefore, these are indicated by symbols (g), (*l*) and (s) placed after the chemical formulae. These stand for gaseous, liquid and solid states respectively. However, sometimes confusion arises in representing the physical state of water formed as a chemical reaction. This can be understood by the following reaction at 25°C :

$$H_2(g) + \frac{1}{2}O_2(g) \rightarrow H_2O(l) + 286.2 \text{ kJ}$$

In the above reaction, since experiment is carried out at 25°C, water is obtained in the liquid state.

If the above reaction is carried out at a temperature a little above 100°C under atmospheric pressure, the water produced will remain in the vapour state. The heat evolved in the reaction is 245.5 kJ. The thermochemical equation will now be written as follows :

$$H_2(g) + \frac{1}{2}O_2(g) \rightarrow H_2O(l) + 245.5 \text{ kJ.}$$

(2) Allotropic Forms of Substances

As different allotropic forms of a substance are associated with different amounts of energy, it means that in case of elements showing allotropy, the name of the allotropic form must be mentioned in thermochemical equation. For example, in a thermochemical equation involving sulphur, we must mention the allotropic forms, rhombic, monoclinic, etc. Thus, in a thermochemical equation, rhombic and monoclinic varieties of sulphur are represented as S_R (rhombic) and SM (monoclinic) respectively.

(3) Reaction in Solutions

When a reaction takes place in a solution, the solvent must be indicated. For example, the symbol (aq) placed after the formula of a substance, reveals that substance is present in large excess of water to ensure that there occurs no heat change on adding further quantity of water.

(4) Heat Changes Involved in Reactions

Two different conventions are used to represent the heat changes of the chemical processes :

(a) *Old Convention :* According to this convention, the heat evolved or adsorbed in a chemical reaction is represented as a part of the equation on the product side. A positive sign (+) is used if heat is given out and negative sign (–) is used if heat is absorbed in a chemical reaction. Thus,

$$A + B \rightarrow C + D + x \text{ kJ}$$

$$A + B \rightarrow C + D - x \text{ kJ}$$

This convention is out of date these days.

(b) *Modern Convention :* This convention is based on thermodynamics. The heat evolved or absorbed in a process may be represented by ΔH, the difference in the total heat contents of the products and reactants. It is not written as a part of equation but is written separately. If heat is absorbed in a chemical reaction, + ΔH is employed and if heat is evolved. Then –ΔH is employed. For example, the burning of carbon in oxygen to form carbon dioxide was previous represented as

$$C_{graphite} + O_2(g) \rightarrow CO_2(g) + 393.5 \text{ kJ}$$

At present it is represented as (Old convention)

$$C_{graphite} + O_2(g) \rightarrow CO_2(g);$$

$$\Delta H = -393.5 \text{ kJ}$$

(5) Temperature and pressure of Substances

The heat absorbed or evolved in a chemical reaction has been found to depend upon the temperature and pressure at which the reaction is being carried out. These two quantities are, therefore, also in the thermochemical equation.

If a reaction is carried out under standard conditions *i.e.*, 25°C or 298 K and 1 atm pressure, the heat change is indicated as ΔH°. The superscript' indicates that pressure is 1 atmosphere and reaction in the standard state.

HEAT OF REACTION OR ENTHALPY OF REACTION

When a chemical reaction is accompanied by evolution of heat, it is termed as exothermic and if it is accompanied by absorption of heat, it is termed as endothernric. *The amount of heat evolved or absorbed during a chemical reaction is called the heat of reaction or enthalpy of reaction.*

From the definition of enthalpy, there always occurs a definite enthalpy change, it follows that it is ΔH, whenever a system undergoes a change of state or a chemical reaction.

Suppose we consider a chemical system having two reactants A and B in the initial state.

Suppose it undergoes a chemical reaction at a constant temperature and pressure, thereby forming the products C and D.

$$A + B \rightarrow C + D$$

Suppose H_1 denotes the enthalpy of the systems in the initial state (*i.e.*, enthalpy of reactants A and B) and H, the enthalpy of the system in the final state at the same temperature and pressure (*i.e.*, enthalpy of products C and D). Then,

$$\Delta H - H_2 - H_1$$

Thus, *at constant pressure and temperature the heat of reaction or enthalpy of reaction is regarded as a measure of the enthalpy chaise, ΔH. In simple words,*

"The heat of reaction or enthalpy of reaction at a given temperature and pressure is the difference between the heat contents of the products and the reactants when the number of gram molecules of substances indicated by the chemical reaction have completely reacted.

Mathematically, it may be put as

ΔH = Heat contents of the products – Heat contents of the reactants.

= Heat contents of 1 mole of C and 1 mole of D – Heat contents of 1 mole of A and 1 mole of B.

ΔH is positive for endothermic reactions and negative for exothermic reaction. For example, the thermochemical equation

$$H_2(g) + \frac{1}{2}O_2(g) \rightarrow 2H_2O(l);\ \Delta H = -285.83\ U$$

indicates that when one mole of gaseous hydrogen combines with half a mole of gaseous oxygen at 298 K at one atmosphere pressure to form 1 mole of liquid water at the same temperature and pressure, 285.83 kJ of heat is given out to the surrounding. It is further assumed that the reaction is fast and proceeds to completion with no side products.

Enthalpy of a Reaction at Constant Pressure and at Constant Volume : The enthalpy of a reaction depends upon the condition under which the reaction is carried out. There are two general conditions under which thermochemical measurements are made, namely :

(i) at constant volume, and

(ii) at constant pressure.

The magnitude of the anthalpy changes in these two conditions are, in general, different.

Heat of reaction or enthalpy of reaction at constant volume (ΔE) : When a reaction is being carried out at a constant volume, the heat change is called the heat of reaction or enthalpy of reaction at constant volume. Example is the combustion of a substance in a bomb calorimeter.

When a reaction is carried out under constant volume by carrying out the reaction in a closed and rigid container, no work is involved and from the first law of thermodynamics, we have

$$\Delta E = qv$$
$$= \text{Enthalpy of the reaction at constant volume.}$$

Thus, the enthalpy of the reaction at constant volume is exactly equal to the change in the internal energy ΔE of the reacting system. In other words, the thermal change that occurs in a chemical reaction is only due to the difference in the sum of the internal energy of the products and the sum of the internal energy of the reactants.

$$\Delta E = \Sigma E_{products} - \Sigma E_{reactants} \quad ...(1)$$

The importance of Eq. (1) is that the amount of heat absorbed at constant volume can be identified with the change in the thermodynamic quantity.

Heat of reaction or Enthalpy of reaction at constant pressure ΔH : When pressure is kept constant throughout the reaction, the heat change is then termed as heat of reaction at constant pressure.

At constant pressure, the system is either kept open to the atmosphere or confined with in a vessel on which a constant external pressure is exerted. Under these conditions, the volume of system changes. The thermal change at constant pressure not only involves the change in internal energy of the system but also the work done either in expansion or in contraction of the system. Then, from the first law of thermodynamics, we have

$$q_P = \Delta E + w$$

If w is only pressure volume work, then

$$q_P = \Delta H + P\Delta V = (\Sigma E_P - \Sigma E_R) + P(V_P - V_R)$$

$$= (\Sigma E_P - PV_P) - (\Sigma E_R + PV_R) \qquad ...(2)$$

where the subscripts P and R refer to the products and the reactants respectively.

But $$H = E + PV \qquad \text{(By definition)}$$

Therefore, Eq. (2) becomes as follows :

$$q_P = \Sigma H_P - \Sigma H_R = \Delta H \qquad ...(3)$$

From the above equation it follows that at constant pressure, the heat of the reaction is exactly equal to the enthalpy change. ΔH. of the reacting system.

Relationship between ΔH and ΔE : In the case of reactions involving solids and liquids, the difference between ΔH and ΔE is almost negligible. Thus, for solids and liquids, $\Delta H \approx \Delta E$. On the other hand, ΔH may not be equal to ΔE in reactions which involve gases. Suppose consider the following reactions :

$$2NH_3(g) \rightarrow N_2(g) + 3H_2(g)$$

2 moles 1 mole 3 moles

$$2H\text{–}2(g) + O_2(g) \rightarrow 2H_2O(g)$$

2 moles 2 moles 1 mole

When these reactions are carried out at constant volume, no work will be done by or on the system as there is neither expansion nor contraction. However, if the reactions are carried out at constant pressure, then the work will be performed by the system in the first case as there occurs an increase in volume

$$(2NH_3 + N_2 + 3H_2)$$

and upon the system in the second case

$$(2H_2 + O_2 \rightarrow 2H_2O)$$

as there occurs a decrease in volume. Consequently, the heat of reaction at constant volume (ΔE) will be more than heat of reaction at constant pressure (ΔH) when contraction occurs during the course of reaction whereas ΔE will be less than ΔH if expansion occurs during the course of reaction. Now consider the following gaseous reaction :

$$aA + bB + ... \rightleftharpoons cC + dD + ...$$

Suppose n_1 = No. of moles of gaseous reactants, *i.e.*, a + b +...

n_2 = No. of moles of gaseous products, *i.e.*, c + d +...

$\Delta n(g) = (c + d + ...) - (a + b + ...) = (n_2 - n_1)$,

V_1 = volume of the gaseous reactants,

P = the external pressure of the gas, and

V_2 = volume of the gaseous products,

T = temperature of the system on absolute scale,

For an ideal gas,

$$PV = nRT$$

$\therefore PV_1 = nRT$ and $PV_2 = n_2RT$

Now, work done by the gas in expansion or contraction from volume V_1 to V_2 in the reaction

$$aA + bB + ... \rightleftharpoons cC + dD + ...$$

is given by

w = pressure × change in volume

$= P(V_2 - V_1) = PV_2 - PV_1$

$= n_2RT - n_1Rt\ (n_2 - n_1) = \Delta n(g)RT$ ...(4)

Thus, heat of reaction at constant pressure is given by,

ΔH = heat of reaction at constant volume + work done by or on the gas during change of volume V_1 to V_2

$= \Delta E + \Delta n(g)RT$ ...(5)

Equation (5) can be used in calculating the value of ΔH from ΔE or vice-versa. A positive value of ΔH or ΔE shows that heat is absorbed while a negative value of ΔH or ΔE shows that heat is evolved or given out by the system to the surroundings,

Case I

When $n_2 > n_1$, $\Delta n(g)$ is positive. Therefore,

$$\Delta H > \Delta E$$

Case II

When $n_2 > n_1$, $\Delta n(g)$ is negative.

Therefore, $\Delta H < \Delta E$

Case III

When there is no change in volume, $n_2 = n_1$. Therefore,

$$\Delta n(g) = 0,\ \Delta H = \Delta E$$

FACTORS ON WHICH HEAT OF REACTION OR ENTHALPY OF REACTION DEPENDS

The important factors which influence the heat of a reaction are given as follows :

(i) *Physical States or Reactants or Products :* For example when red and yellow phosphorus burn in oxygen to form P_2O_5, different values of heats of reactions are obtained.

$$4P_{yellow} + 5O_2(g) \rightarrow P_4O_{10};$$
$$\Delta H = -3024 \text{ kJ}$$
$$4P_{red} + 5O_2(g) \rightarrow P_4O_{10};$$
$$\Delta H = -3026 \text{ kJ}$$

Thus, there is a difference in the heats of reactions of the two allotropic forms of phosphorus.

(ii) *Temperature :* Heat of reaction has been found to depend upon the temperature at which reaction is occurring. The variation of the heat of reaction which temperature can be ascertained by using Kirchhoff's equation

(iii) *Heats of solution of Reactants and Products :* It is found that different values of heats of reaction are obtained when reactions are carried out with the anhydrous reactants or in solution. For example,

$$H_2S(g) + I_2(g) \rightarrow 2HI(g) + S \; ; \; \Delta H = -72.24 \text{ kJ}$$

$$\underset{\text{(Saturated) solution}}{H_2S(g) + I_2} \rightarrow \underset{\text{Solution}}{2HI + S} \; ; \quad \Delta H = -91.98 \text{ kJ}$$

EXOTHERMIC AND ENDOTHERMIC REACTIONS

(a) *Exothermic Reactions* : A reaction which proceeds with the evolution of heat is referred to as an exothermic reaction. Some examples of exothermic reactions are as follows :

$$2Zn(s) + O_2(g) \rightarrow 2ZnO(s) + 693.8 \text{ kJ} \quad ...(1)$$

$$C(s) + O_2(g) \rightarrow CO_2(g) + 393.5 \text{ kJ} \quad ...(2)$$

$$C_2H_2(g) + 2H_2(g) \rightarrow C_2H_6(g) + 314.0 \text{ kJ} \quad ...(3)$$

In all the above reactions, heat is evolved and therefore these are examples of exothermic reactions.

If an exothermic reaction is taking place at constant temperature and constant volume, the energy change is expressed in terms of internal energy changes, *i.e.*, ΔE. Tηυσ,

ΔE = Internal energy of product - Internal energy of reactants = $E_P - E_R$

In exothermic reactions, the total internal energy of the reactants is more than that of products and therefore the excess of internal energy, ΔE, will l appear as heat of reaction or in other words ΔE is negative for exothermic reactions. Thus, equations (1), (2) and (3) in terms of ΔE can be written as :

$$2Zn(s) + O_2(g) \rightarrow 2ZnO(S) \; ; \; \Delta E = -693.8 \text{ kJ}$$

$$C(s) + O_2(g) \rightarrow CO_2(g) \; ; \; \Delta E = -393.5 \text{ kJ}$$

$$C_2H_2(g) + 2H_2(g) \rightarrow C_2H_6(g) \; ; \; \Delta E = -314.0 \text{ kJ.}$$

If an exothermic reaction is taking place at constant temperature and constant pressure, the energy change is expressed in terms of enthalpy change, ΔH, *i.e.*,

ΔH = Enthalpy of products– Enthalpy of reactants = $H_P - H_R$

In exothermic reactions, the total heat content of the reactants is more than that of products and, therefore, the excess of heat constant will appear as heat of reaction. In other words ΔH like ΔE is negative. Thus, equations (1), (2) and (3) in terms of ΔH can be written as :

$$2Zn(s) + O_2(g) \rightarrow 2ZnO(S) ; \Delta H = -693.8 \text{ kJ}$$

$$C(s) + O_2(g) \rightarrow CO_2(g) ; \Delta H = -393.5 \text{ kJ}$$

$$C_2H_2(g) + 2H_2(g) \rightarrow C_2H_6(g) ; \Delta H = -314.0 \text{ kJ}$$

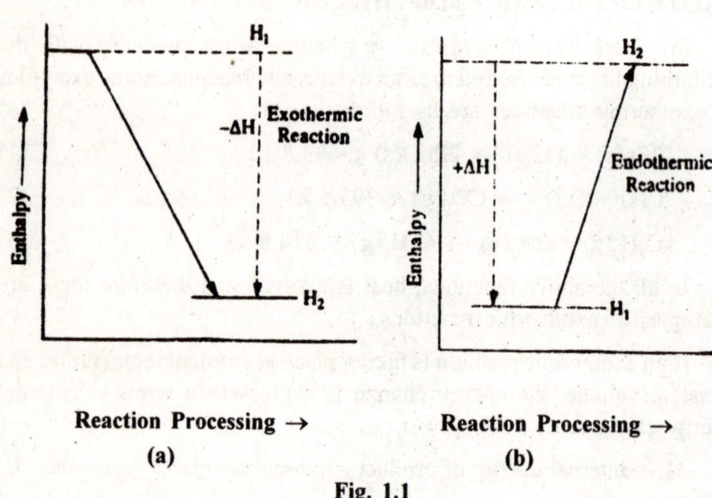

Fig. 1.1

As ΔH is negative for exothermic reactions, this is illustrated in Fig. 1.1 (a).

(b) *Endothermic Reactions* : A reaction which proceeds with the absorption of heat is referred to as an endothermic reaction. Some examples of endothermic reaction are as follows :

$$2HgO(s) + 180.4 \text{ kJ} \rightarrow 2Hg(l) + O_2(g) \quad ...(1)$$

$$H_2O(g) + C(s) + 131.2 \text{ kJ} \rightarrow CO(g) + H_2(g) \quad ...(2)$$

$$N_2(g) + O_2(g) + 180.5 \text{ kJ} \rightarrow 2NO(g) \quad ...(3)$$

If endothermic reactions are taking place at constant temperature and volume, the total internal energy of the reactants (E_R) is less than the total internal energy of the products (E_P), *i.e.*, ΔE is positive. Thus, equations (1), (2) and (3) can be put as :

$$2HgO(s) \rightarrow Hg(l) + O_2(g); DH = 180.4 \text{ kJ}$$

$$H_2O(g) + C(s) \rightarrow CO(g) + H_2(g); DH = 131.2 \text{ kJ}$$

$$N_2(g) + O_2(g) \rightarrow 2NO(g); DE = 180.5 \text{ kJ.}$$

If endothermic reactions are carried out at constant pressure and constant temperature, then the total heat content of products is more than the total heat content of reactants. In other words, DH is positive. It is represented in Fig. 1.1 (b). In terms of DH, equations (1) (2) and (3) may be put as:

$$2HgO(g) \rightarrow 2Hg(l) + O_2(g); \; DE = 180.4 \text{ kJ}$$

$$H_2O(g) + C(s) \rightarrow CO(g) + H_2(g); \; DE = 131.2 \text{ kJ}$$

$$N_2(g) + O_2(g) \rightarrow 2NO(g); \; DH = 180.5 \text{ kJ}$$

Melting of solids and vaporization of liquids are endothermic reactions. Synthesis of proteins in living systems is an endothermic process.

(c) Differences between Exothermic and Endothermic Reactions.

The differences are given in tabular from :

Table 1.1

Exothermic Reactions	*Endothermic Reactions*
1. Heat is given to the surroundings.	1. Heat is absorbed from the surroundings.
2. The internal energy of reactants is more than that of products, *i.e.*, ΔE is – ve.	2. The internal energy of products is more than that of reactants, *i.e.*, ΔE is + ve.
3. The enthalpy of reactants is more than that of Products, *i.e.*, ΔH is –ve.	3. The enthalpy of reactants is less than that of Products, i.e, ΔH is +ve.

STANDARD ENTHALPY CHANGES

As the enthalpy of a reaction depends on temperature, a standard state is chosen to compare the enthalpies or reactions of different substances. A convenient standard state for a substance is the most stable state of aggregation under 1 atmosphere and at the specified temperature, generally at 25°C or 298 K.

For pure solids, liquids and ideal gases, the standard state corresponds to the state of the substances at 1 atm. Pressure and specified temperature.

For the dissolved substances, the standard state of the solute is that concentration which gives unit activity.

The standard enthalpy of a reaction at a temperature T and at a pressure of 1 atm is denoted by $\Delta H^o(T)$ and is the value of $\Sigma \Delta H_{products} - \Sigma \Delta H_{reactants}$, at that temperature when all the substances (or elements) in the reaction are in their standard states.

The enthalpy of every element in its standard state is arbitrarily given a zero value. For example, the enthalpy of formation of a mole of $CO_2(g)$ from its elements in the standard state is – 393.5 kg mol^{-1}, *i.e.*,

$$C(graphite) + O_2(g) \rightarrow CO_2(g);\ \Delta H^o = -\ 393.5\ kJ.$$

The value of ΔH^o_{298K} is given as follows:

$$\Delta H^o_{298K} = H^oCO_2(g) - [H^oO_2(g) + H^o_{Cgrphite}]$$

In the above equation, H° refers to the standard enthalpy per mol.

As the enthalpies are in their standard states, their heat contents by convention are zero, *i.e.*,

$$H^o_{Cgrphite} = H^o_{O2(g)} = 0.$$

Hence under these conditions,

$$\Delta H^o_{298K} = H^o_{CO2(g)} = -\ 393.5\ kJ.$$

Thus, the standard enthalpy of any compound is the enthalpy of the reaction by which it is formed from its elements in the standard state.

VARIOUS TYPES OF ENTHALPY CHANGES

Enthalpy of a reaction is given different names depending upon the types of reactions. Some of those are described below.

1. Heat or Enthalpy of Formation

The enthalpy of formation of a compound at the given temperature may be defined as the enthalpy change when one mole of a compound (in its standard state) is formed its elements (in their standard states). It is usually denoted by ΔH^o_f.

In the above definition, the standard state of any substance is taken as its natural state at 298 K under one atmosphere pressure. Let us consider the following reactions :

$$C(graphite) + O_2(g) \rightarrow CO_2(g);\ \Delta H = \Delta H^o_f = -\ 393.5\ kJ$$

$$C(graphite) + 2H_2(g) \rightarrow CH_4(g);\ \Delta H = \Delta H^o_f = -\ 74.8\ kJ$$

Sometimes, the enthalpy of combustion and enthalpy of formation relate to the same reaction ; for instance, the enthalpy of formation of carbon dioxide is the same as the enthalpy of combustion of graphite.

The compound with negative enthalpy of formations called exothermic compound whereas the compound with positive enthalpy of formations known as endothermic compound. Thus, the formation of water is an exothermic compound whereas the formation of hydrogen iodide is an endothermic compound.

$$H_2(g) + 1/2O_2(g) \rightarrow H_2O(g);\ \Delta H^o_f = -\ 245.5\ kJ$$

$$1/2H_2(g) + 1/2I_2(s) \rightarrow HI(g);\ \Delta H^o_f = +\ 25.94\ kJ$$

The negative enthalpy of formation of a compound means that its heat content is less than that of its elements and hence it is more stable than its elements. The reason is for that the compound with a lower enthalpy (or energy) must be more stable. Therefore, the chemical bonds in H_2O are stronger than those in H_2 or O_2. On the other hand, the positive enthalpy of formation of a compound means that its enthalpy of content is more than that of its elements and hence it is less stable than its elements. Due to this reason, H – I bond is weaker than H – H bond in hydrogen or I – I bond in iodine.

The enthalpy of formation of a compound may be obtained either by direct measurement of ΔH for the reaction or from enthalpies of reactions involving the compound.

2. Heat or Enthalpy of Combustion

The enthalpy of combustion of a substance at a given temperature may be defined as the enthalpy change accompanying the complete combustion of one mole of the substance in excess of oxygen or air at that temperature.

The words "complete combustion" in the above definition must be emphasized because it is possible to burn a compound to yield both CO and CO_2. In order to avoid this confusion. The combustion is carried out in a large excess of oxygen or air to yield carbon dioxide and water as the final products. For example, heat of combustion relates to the reaction :

$$C_2H_2(g) + 21/2O_2(g) \rightarrow 2CO_2(g) + H_2O(g);\ \Delta H^o = -\ 1258\ kJ$$

and not to the partial oxidation as follows :

$$C_2H_2(g) + 11/2O_2(g) \rightarrow 2CO(g) + H_2O(g); \Delta H^o = -691 kJ.$$

The heat of combustion depends upon the physical state of the substances involved. For example, the heat of combustion of hydrogen at 298 K is –285.9 kJ mol^{-1}.

In this case only liquid water is obtained. On the other hand if the combustion is taking place at 313 K or above, the water vapour in this case cannot exist in the liquid state but evaporates to form vapours. The enthalpy of combustion is that case will be –245.5 kJ.

$$H_2(g) + 1/2O_2(g) \rightarrow H_2O(1); \Delta H^o_{298k} = -285.19 \text{ kJ at } 298 \text{ k}$$

$$H_2(g) + 1/2O_2(g) \rightarrow H_2O(g); \Delta H^o_{373k} = -245.5 \text{ kJ at } 373$$

The enthalpy of combustion at 373 K is lower as compared at 298 K. The reason for this is that small quantity of energy (+ 40 .44 kJ) is used up to convert liquid water into its vapours.

$$H_2O(l) \rightarrow H_2O(g); \Delta H^o - +40.44 \text{ kJ}.$$

The enthalpy of combustion of methane is

$$CH_4(g) + 2O_2(g) \rightarrow CO_2(g) + 2H_2O(l); \Delta H^o = -890.40 \text{ kJ}.$$

Enthalpy of combustion may be used to calculate the enthalpy of formation of a compound provided the enthalpy of formation of other combustion products are known.

For example, it is possible to calculate enthalpy of formation of gaseous methane provided the enthalpies of formation of $CO_2(g)$ and $H_2O(l)$ are known.

$$\Delta H^o = -890.40 = \Delta H^o_{CO_2(g)} + 2\Delta H^o_{H_2O(l)} - \Delta H^o_{CH_4(g)}$$

$$\text{or } \Delta H^o_{CH_4(g)} = 890.40 + \Delta H^o_{CO_2(g)} + 2\Delta H^o_{H_2O(l)}$$

$$= 890.40 + (-393.50) + 2(-285.83) = -74.76 \text{ kJ mol}^{-1}.$$

Application of Heat Combustion Date : Main applications are :

(i) Calorific value of fuels

(ii) Food values of diets

(iii) Maximum flame temperature

(iv) Calculation of heal of formation

(v) Calculation of heat of reaction.

3. Heat or Enthalpy of Hydrogenation

It may be defined as the enthalpy change when a mole of an unsaturated organic compound is fully hydrogenated. The enthalpies of hydrogenation of ethylene and benzene are represented as follows .

$$CH_2 = CH_2(g) + H_2(g) \rightarrow CH_3 - CH_3(g);$$

$$\Delta H^o = -\ 132.5 \text{ kJ}$$

$$C_6H_6(l) + 3H_2(g) \rightarrow C_6H_{12}(l);$$

$$\Delta H^o = -\ 246 \text{ kJ}$$

From the above thermochemical equation, it is to be noted that enthalpy of hydrogenation of benzene is not three times that of enthylene one may expect this due to the presence of three double bonds in benzene and one double bond in enthylene.

The enthalpy of hydrogenation of benzene is less be 151.5kJ mol^{-1} which is attributed to the stabilisation on benzene due to resonance. This method may sometimes be used as the characteristic of stabilisation of aromatic molecules due to resonance.

4. Heat or Enthalpy of Neutralization

The enthalpy of neutralization of an acid at a given temperature may be defined as *the enthalpy change accompanying the neutralization of one gram equivalent of a base by an acid in dilute solution at that temperature.*

For example, the enthalpy of neutralization one of hydrochloric acid by sodium hydroxide in dilute solution at 298 K is – 57.32 kJ.

$$NaOH + HCI \rightarrow NaCI + H_2O$$

As neutralization is the reaction between $H^+(aq)$ form the base, one can write the above equation as :

$$H^+(aq)+CI^-(aq)+Na^+(aq)+OH^-(aq) \rightarrow H_2O(I)+Na^+(aq)+CI^-(aq)$$

or $$H^+(aq) + OH^-(aq) \rightarrow H_2O(I)$$

From this it follows that enthalpy of neutralization should be independent of the nature of acid and the base.

This has been found to be true in the case of neutralization of strong acids by strong bases. This is evident form the data given in Table 2.1.

Table 1.2 : Enthalpies on Neutralization of Strong Acids by Strong Bases at 298 K

Acid	*Base*	*Enthalpy of Neutralization (kJ per gram equivalent)*
HCl	NaOH	–57.32
HCl	KOH	–57.45
HCl	LiOH	–57.32
HNO_3	NaOH	–57.28

5. Heat or Enthalpy of Ionisation

For a weak acid or a weak base, the enthalpy of neutralisation is not the same but less than –57.32 kJ. It is because of the fact that these acids or bases are not completely ionised in solution. Some of the heat is consumed in ionising these acids and bases. The heat required to ionise the weak acid or the weak base is known as the enthalpy of ionisation.

It is possible to calculate the enthalpy of ionisation of a weak acid from the enthalpies of neutralisation of strong acid and weak acid by strong bases. For example, in the neutralisation of acetic acid by sodium hydroxide, the enthalpy of neutralisation is – 55.43 kJ mol^{-1}. The enthalpy of ionisation of acetic acid will therefore be equal to +1.89 kJ mol^{-1}. this can seen as follows :

(i) $CH_3COOH(aq) + OH^-(aq) \rightarrow CH_3COO^-(aq) + H_2O(l)$;

$\Delta H^o = -55.43$ kJ

(ii) $H^+(aq) + OH^-(aq) \rightarrow H_2O(l)$; $\Delta H = -57.32$ kJ

(iii) $CH_3COOH(aq) \rightarrow CH_3COO^-(aq) + H^+(aq)$;

ΔH^o Ionisation = ?

On adding Eqs. (ii) and (iii), we obtain

$CH_3COOH(aq) + OH^-(aq) \rightarrow CH_3COO^-(aq) + H_2O(l)$

or $\Delta H^o = -57.32 + \Delta H^o$ Ionisation

But from Eq. (i), we have

$CH_3COOH(aq) + OH^-(aq) \rightarrow CH_3COO^-(aq) + H_2O(l)$

or $\Delta H^o = -55.43$ kJ mol^{-1}

Hence ΔH^o Ionisation = $-55.43 + 57.32 = +1.89$ kJ mol.

6. Heat or Enthalpy of Solution

When a solution is dissolved in a solvent, heat may be either liberated or absorbed. If the heat is absorbed the solution becomes cooler and ΔH is given a positive sign. On the other hand if the heat is evolved, the solution gets warmer and ΔH is given a negative sign.

The enthalpy change per mole of a solute dissolved varies to a considerable extent with dilution and to some extent with the initial temperature of the solvent. Therefore, it becomes essential to express the enthalpy change with reference to the concentration of the solution.

"The enthalpy of solution may be defined as the enthalpy change when one mole of the solute is dissolved in a specified amount of solvent.''

The enthalpy of solution can be explained by considering the dissolution of HCl (g) in water at 298 K:

$$HCl\ (g) + 25H_2O(l) \rightarrow HCl.25\ H_2O;\ \Delta H^\circ = -69.5\ kJ$$

$$HCl\ (g) + 40H_2O(l) \rightarrow HCl.40\ H_2O;\ \Delta H^\circ = -73.18\ kJ$$

$$HCl\ (g) + 200H_2O(l) \rightarrow HCl.200H_2O;\ \Delta H^\circ = -74.40\ kJ$$

$$HCl\ (g) + \infty\ H_2O(l) \rightarrow HCl.(aq);\ \Delta H^\circ = -75.17\ kJ$$

From the above equation it means that when one mole of HCl is dissolved in 25 moles of water, 69.5 kJ of heat are evolved. But if the solution is further diluted to such an extent that further dilution would cause no additional heat change, then the equation may be written as :

$$HCl\ (g) + \infty\ H_2O(l) \rightarrow HCl.(aq);\ \Delta H^\circ = -75.17\ kJ\ mol^{-1}$$

where aq. represents a large volume of water. The value of ΔH° correspondents to this reaction is called the enthalpy of solution at infinite dilution. It is the maximum quantity of heat released in the formation of a solution of one mole of HCl(g) in excess of water.

For greater accuracy, two types of heat of solution have been recognised. There are :

(a) *Integral Heat of Enthalpy of Solution :* Integral heat of solution at a particular concentration is the enthalpy change when 1 mole of solute is dissolved in pure solvent to produce a solution of the given concentration. Thus, if 1 mole of solute is dissolved in 500g of water the heat change gives the value of integral heat of solution at the concentration of 2 molal.

While recording integral enthalpies of solution, the usual practice is to state the amount of solvent in which one mole of solute is dissolved. For example :

$$HCl\ (g) + 10H_2O(l) \rightarrow HCl.(10\ H_2O);\ \Delta H^o \doteq -69.5\ kJ.$$

indicates that when one mole of hydrogen chloride gas is dissolved in 10 moles of water, there is an evolutions 65.5 kJ of heat energy.

According to Hess's law, the integral enthalpy of solution is equal to the difference between integral enthalpies of the solution at the two concentrations. For example, if to a solution of 1 mole of hydrogen chloride in 40 moles of water, enough water is added such that

$$HCl.40\ H_2O + aq \rightarrow HCl.(aq)$$

the integral enthalpy change can be calculated as follows :

$$HCl(g) + aq \rightarrow HCl\ ; \quad \Delta H^o = -\ 75.17\ kJ$$

$$HCl(g)\ 40\ H_2O \rightarrow HCl.\ 40H_2O;\ \ \Delta H^o = -\ 73.18\ kJ$$

On subtracting, we have

$$HCl.40\ H_2O + aq \rightarrow HCl.(aq); \quad \Delta H^o = -\ 1.99\ kJ$$

If ΔH is plotted against number of solute per 1000 g solvent, the resultant curve is shown in Fig.1.2. Initially, ΔH increases almost linearly with m but later the increase is not as fast as m. Finally, it reaches a constant value when the solution becomes saturated with respect to solute. If the observers value of enthalpy change (ΔH) is divided by the

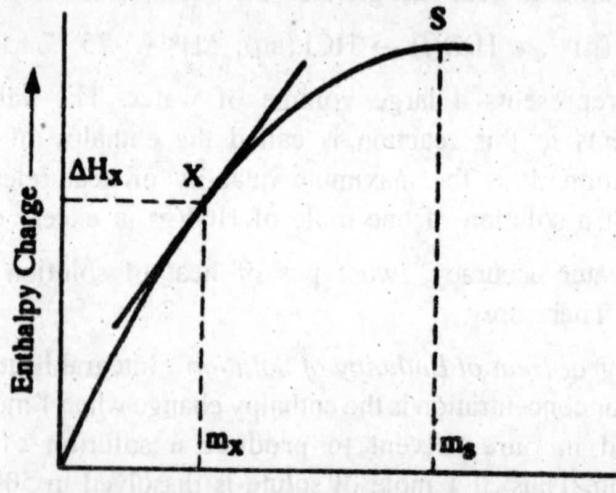

Fig .1.2 : Variation of enthalpy change when the amount m of solute is dissolved in 1 kg.

amount of solute (m) that is dissolved to form a solution of a particular concentration, we get the quantity $\Delta H/m$, which is nothing but the integral enthalpy of solution at the given concentration.

Integral enthalpies of solutions, enthalpies of dilutions and enthalpies of reactions in solutions can be evaluated from the enthalpies of formation in solution. In these calculations, the enthalpy of formation of water is neglected if the same number of moles of water appears on both sides of the balanced thermochemical equation.

(b) *Differential Heat or Enthalpy of Solution :* The differential heat of solution, $DH_{D.s}$ may be defined as the heat of solution of a mole of solute in such a large quantity of solution of known concentration that addition of 1 more mole of solution does not change the concentration appreciably. If on dissolving dm mole of the solute in a finite amount of the solution the heat change is dq, then differential heat of solution will be

$$\Delta H_{D.s} = \delta q/\delta m$$

From the shape of the curve of Fig. 1, it can be seen that the value of the slope will depend upon the concentration of the solution. Thus, the differential enthalpy of a solution besides depending to T and p, will also depend on the amounts of solvent n_1 and solute n_2 that are present in the given solution *i.e.*,

$$d\,(\Delta H)/dm = f(T,\ p\ n_1,\ n_2)$$

Thus, for this reason the concentration of the solution is mentioned while defining the differential enthalpy of solution. The following conclusions may be drawn from the shape of the curve between ΔH and m as shown in Fig. 1.

(i) For smaller value of m, the curve is almost linear; thus its slope will have a constant value and will be equal to DH/m. As the latter represents the integral heat of solution, it means that the differential and integral heats of solution are essentially equal for very dilute solutions.

(ii) For higher values on m, the curve is not linear. The reason is that ΔH does not increase as fast as m; thus the slope of the curve decreases as the value of m increases. In other words, the differential heat of solution decreases as the concentration of the solution increases, and becomes zero when the solution gets saturated.

The differential enthalpy of solution can be obtained by plotting ΔH of the solution at various concentrations against the number of moles of solvent associated with a definite quantity of solute, and finding the slope of the curve at a point corresponding to any particular concentration. The value of the slope at a given concentration gives the differential enthalpy of solution at that concentration.

The differential heats of solution and dilution are essentially the partial molar heat of solute and solvent, respectively. Consider a solution having n_1 moles of solvent and n_2 moles of solute. In general,

$$\Delta H = f(T, p\ n_1, n_2)$$

The differential of ΔH is given as follows :

$$d(\Delta H) = \left(\frac{\partial(\Delta H)}{\partial T}\right)_{P,\ n_1,\ n_2} dT + \left(\frac{\partial(\Delta H)}{\partial P}\right)_{T,\ n_1,\ n_2} dP + \left(\frac{\partial(\Delta H)}{\partial n_1}\right)_{T,P,\ n_2} dn_1 + \left(\frac{\partial(\Delta H)}{\partial n_2}\right)_{T,P,\ n_1} dn_2$$

At constant temperature and pressure, the above equation becomes as follows :

$$d(\Delta H) = \left(\frac{\partial(\Delta H)}{\partial n_1}\right)_{T,P,\ n_2} dn_1 + \left(\frac{\partial(\Delta H)}{\partial n_2}\right)_{T,P,\ n_1} dn_2$$

$$\text{or } d(\Delta H) = \left(\Delta\overline{H}_1\right) dn_1 + \left(\Delta\overline{H}_2\right) dn_2$$

where ΔH_1 and ΔH_2 are termed as partial molar values of heat of solvent and solvent and solute, respectively. It can be seen that ΔH_2 is equal to the slope d ($\Delta H/dn_2$) of a varve in which the property ΔH is plotted as a function of n_2 for a series of solutions in which n_1 is kept constant. When it is compared with the definition given at the beginning it shows that the differential heat of solution is really the partial molar heat of solution of solute, *i.e.*, ΔH_2. Similarly, the differential heat of dilution is the corresponding partial molar property of the solvent, *i.e.*, $\Delta\overline{H}_1$.

Direct experimental measurement of differential heat of solution is not possible, it can, however, be determined indirectly from the knowledge of integral heat of solution

$$\Delta H_{D.S} = d\ (m\ \Delta H_{I.S})/dm$$

where $m\Delta H_{I.S}$ is the integral heat of solution of the concentration m molal.

As it is obvious from the definitions, the integral and differential heats of solutions approach each other with decreasing the concentration and become idential when the concentration is zero.

Enthalpy of Dilution : If it known than t the enthalpy of solution per mole of a solute gets varied with the concentration of the solution. Therefore, it follows that if a given solution is diluted on the addition of solvent, there occurs noticeable change of enthalpy. This is termed as the enthalpy of dilution.

Enthalpy of dilution may be defined as the change of enthalpy when a solution having 1 mole of solute is diluted from one concentration to another.

7. Heat or Enthalpy of Hydration

The heat or enthalpy of hydration of salt is the change in heat content ΔH when one mole of an anhydrous substance combines with requisite number of moles of water to form the hydrate. It is obtained from the enthalpies of solutions of hydrous and anhydrous substances. Enthalpy of hydration of copper sulphate can be calculated from the following thermochemical equations :

(i) $CuSO_4.5H_2O + aq \rightarrow CuSO_4\ (aq)$; $\Delta H^o = 11.72$ kJ

(ii) $CuSO_4(s) + 5H_2O \rightarrow CuSO_45H_2O\ (s)$; $\Delta H^o = x$ kJ

(iii) $CuSO_4(s) + 5H_2O\ (l) + aq \rightarrow CuSO_4\ (aq)$; $\Delta H^o = -66.5$ kJ

From the above equations, we get

$$CuSO_4\ (s) + 5H_2O\ (l) \rightarrow CuSO_4\ 5H_2O(s);\ \Delta H^o = -78.22\ kJ.$$

8. Heat or Enthalpy of Transition

It may be defined as the enthalpy change when one mole of one allotropic form changes to another. It includes transition from solid to liquid (fusion), liquid to vapour (vaporisation), solid to vapour (sublimation) and change from one crystalline form to another crystalline form (polymorphic transition). For example, in the transition of sulphur (rhombic) to sulphur (monoclinic), The enthalpy change is 13.14 kJ.

$$S_{rohmbic} \rightarrow S_{monoclinic};\quad \Delta H^o = 13.14\ kJ$$

Another example is

$$C_{graphite} \rightarrow C_{diamond}; \quad \Delta H^o = 1.90 \text{ kJ}$$

The enthalpy of transition in the above case can be obtained from the enthalpies of combustion of $C_{graphite}$ and $C_{diamond}$.

(i) $C_{graphite} + O_2\ (g) \rightarrow CO_2\ (g); \Delta H^o = -\ 393.51$ kJ

(ii) $C_{diamond} + O_2\ (g) \rightarrow CO_2\ (g): \Delta H^o = -\ 395.41$ kJ

On subtracting (ii) from (i), we get

$$C_{graphite} \rightarrow C_{diamond}; \quad \Delta H^o = 1.90 \text{ kJ}$$

9. Heat or Enthalpy of Precipitation

It may be defined at the enthalpy change when one mole of a precipitate is formed upon mixing dilute solutions of relevant electrolytes. Some examples are :

$BaCl_2$ (aq) + Na_2SO (aq) → $BaSO_4$(s) + 2NaCI (aq);

$\Delta H^o = 19.5$ kJ

As the reactants are strong electrolytes, the reaction is essentially ionic in nature and may be written as follows :

Ba^{2+} (aq) $2CI^-$(aq) + $2Na^+$(aq) + SO_4^{2-}(aq)

→ $BaSO_4$(s) + $2Na^+$ (aq) + $2CI^-$(aq), $\Delta H = -\ 19.50$ kJ

As Na^+(aq) and CI^-(aq) ions are common on both sides, the reaction, in fact, involves precipitation of $BaSO_4$(s), *i.e.*,

Ba^{2+} (aq) + SO_4^{2-}(aq) → $BaSO_4$(s); $\Delta H = -\ 19.50$ kJ

Another example is as follows :

$AgNO_3$ (aq) NaBr (aq) → AgBr(s) + $2NaNO_3$ (aq);

$\Delta H^o = -\ 17.78$ kJ

10. Enthalpy of Fusion

The enthalpy of fusion of a substance may be defined as the enthalpy change when 1 mole of the substance is converted from the solid to the liquid phase at its melting point. When 1 mole of ice melts at 273 K, the enthalpy change is approximately 6.6 kJ. This result can be put in the equation form as :

$$H_2O(s) \rightarrow H_2O(I), \quad \Delta H = 6.0 \text{ kJ at } 273 \text{ K}$$

The enthalpies of fusion of molecular solids (O_2, CCI_4, H_2S, etc.,) are low where as the enthalpies of fusion of ionic solids (NaCI, $MgCI_2$)

are large. Thus one may get some idea from the value of fusion, about the nature of the solid and the magnitude of forces acting between the particles constituting the solid.

11. Enthalpy of Vaporization

The enthalpy of vaporization $\Delta(H_V)$ of a liquid may be defined as the enthalpy change accompanying the conversion of 1 mole of the substances form the liquid to the vapour phase at its boiling point. In order to cover liquid into its vapours the substance will absorb energy from the surroundings and thus ΔH will have + ve sign. For example, if water is heated at its boiling point, 373 K, it is converted into vapours by absorbing 40.6 kJ mol^{-1} energy from the surroundings. This means ΔH_V of water is 40.6 kJ mol^{-1}.

$$H_2O(l) \rightarrow H_2O(g); \quad \Delta H_V = +\ 40.6 \text{ kJ mol}^{-1}$$

Similarly, ΔH_V for hydrogen and oxygen are 0.904 and 6.7 kJ Mol–1 respectively. If we compare these values with water we realise that the interaction between water molecules is much stronger than that between hydrogen molecule sin the liquid state but less stronger than that between oxygen molecules in the liquid state. This is quite true because in water there is extensive hydrogen bonding whereas in oxygen and hydrogen there are only weak van der waals forces,

The enthalpy of fusion is very small as compared to the enthalpy of vaporization. The reason for this is that attractive forces in solid and liquid state are much similar (intermolecular distances in solids and liquids are roughly the same) whereas in the gaseous state the attractive forces are much weaker.

12. Enthalpy of Sublimation

The enthalpy of sublimation (ΔH_{sub}) of a substance may be defined as the enthalpy change accompanying the conversion to 1 mole of solid directly into vapour phase at a given temperature below the melting point of the former. Generally, molecular solids like iodine and naphthalene sublime readily. For example, the enthalpy of sublimation of iodine is 62.39 kJ mol^{-1}.

According to Hess's law, the enthalpy of sublimation is equal to sum of enthalpy of fusion (ΔH_f) and enthalpy of vaporization (ΔH_V), *i.e.*,

$$\Delta H_{sub} = \Delta H_f + H_v$$

The above equation is very useful because it can be employed to calculate the enthalpy of sublimation of the enthalpy of fusion and enthalpy of vaporization are known.

13. Enthalpy of Atomization

This is defined as the enthalpy change required to convert one mole of an element from its normal state at 298 K and one atmosphere pressure into free atoms.

Enthalpies of atomization, are difficult to measure (they are usually obtained from spectroscopic data), but the are useful in relation to bond energies. For example :

$$H_2.g = 2H_{atoms} \qquad \Delta H = 433 \text{ kJ} \qquad ...(1)$$

$$O_2.g = 2O_{atoms} \qquad \Delta H = 495 \text{ kJ} \qquad ...(2)$$

$$H_{2g} + 1/2O_{2.g} = H_2O_g \qquad \Delta H = -246 \text{ KJ} \qquad ...(3)$$

Then (3) – (1) – 1/2 × (2) gives :

$$2H_{atoms} + O_{atoms} = H_2O \qquad \Delta H = -926 \text{ kJ}$$

This energy change corresponds to the formation of two oxygen-hydrogen bonds, so that the average bond energy of an oxygen-hydrogen bond in water is 463 kJ.

EXPERIMENTAL DETERMINATION OF ΔE AND ΔHD

Determination of ΔE. If a reaction is allowed to take place at constant temperature and constant volume the internal energy change, ΔE, of the reaction is equal to the heat exchanged with the surroundings. Thus, the external energy change accompanying a chemical reaction can be determined by measuring the heat evolved or absorbed provided the reaction is taking place at constant temperature and constant volume.

Let us now describe the experimental method for determining ΔE. A bomb calorimeter which is mainly used to measure ΔE is shown in Fig. 1.

The calorimeter consists of a strong reaction vessel called bomb calorimeter built of heavy steel. For many reactions, the heat is evolved or absorbed provided the reaction is taking place at constant temperature and constant volume.

The calorimeter consists of a strong reaction vessel called bomb calorimeter built of heavy steel. For many reactions, the constant volume requirement results in large pressure changes. For this reason the calorimeter is built of heavy steel to withstand high pressure. The bomb calorimeter is placed in the centre of a larger vessel filled with a weighed quantity of water. The larger vessel is also provided with a stirrer and sensitive thermometer. The larger vessel is generally a metal container and is surrounded by an insulating material so that no heat is lost to the surroundings. A weighed sample is placed inside a platinum cap and the temperature of water is noted, The reaction is done inside the platinum cap by passing an electric current through a wire embedded into it. The heat evolved or absorbed due to the reaction changes the temperature of the entire apparatus. The temperature of water is again noted. From the change in temperature and the total water equivalent of the calorimeter, ΔE can be calculated.

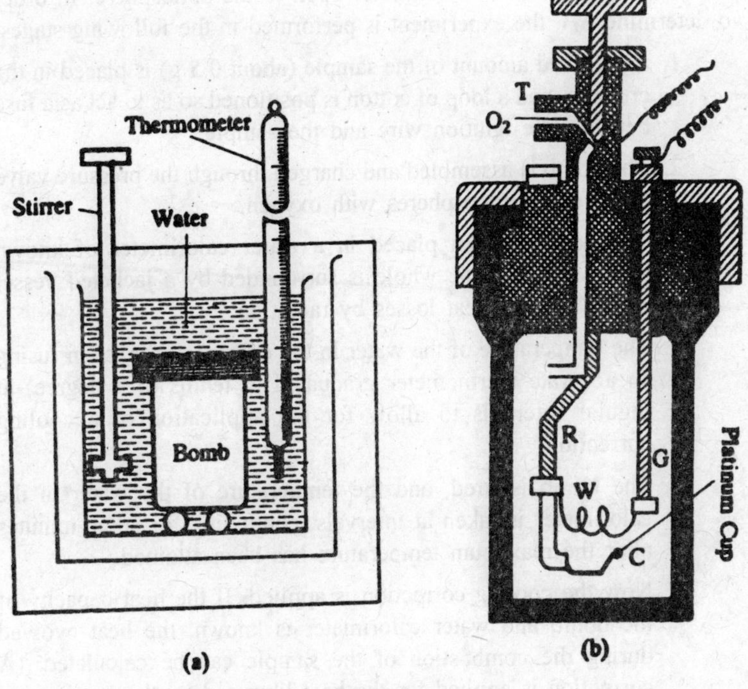

Fig. 1.3

The heat capacity of the calorimeter system is known by burning a known weight of a substance of known heat of combustion. For this purpose, usually benzoic acid of highgrade purity is used. Its heat of combustion which has been very carefully measured is taken as – 3226.7 kJ per mole.

Let the thermal capacity of the calorimeter system including water be z and θ the rise in temperature produced by burning m of the given substance of molecular weight M. Then, evidently, the heat of combustion of the substance will be equal to

$$z \times \theta \times M/m \text{ Joules.}$$

It may be noted that in this case, the volume remains constant. Hence the heat of the reaction gives a measure of q_v.

Determination of ΔH. In order to determine ΔH, a bomb calorimeter is used as in the case of determination of ΔE. But the bomb calorimeter used for the determination of H is not a closed container as used determination of ΔE but a container open to the atmosphere. In order to determine ΔH, the experiment is performed in the following stages:

1. A weighed amount of the sample (about 0.5 g) is placed in the crucible, and a loop of cotton is positioned so as to act as a fuse between the ignition wire and the sample.
2. The bomb is assembled and charged through the pressure valve to 20 or 30 atmospheres with oxygen.
3. The bomb is then placed in a water calorimeter of known equivalent, and the whole is surrounded by a jacketed vessel which reduces heat losses by radiation.
4. The temperature of the water in the calorimeter is taken (using an accurate thermometer graduated in tenths of a degree) at regular intervals to allow for the application of a cooling correction.
5. The bomb is fired, and the temperature of the water in the calorimeter is taken at intervals, continuing for some minutes after the maximum temperature has been attained.
6. Now the cooling correction is applied. If the heat capacity of the bomb and water calorimeter is known, the heat evolved during the combustion of the sample can be calculated. (A correction is applied for the heat liberated by the combustion of the cotton loop).

From the temperature change and the total water equivalent of the calorimeter, ΔH can be calculated.

In order to convert ΔE into ΔH or vice versa, we have to use the equation

$$\Delta H = \Delta E + \Delta nRT$$

KIRCHHOFF'S EQUATION

In 1858 Kirchhoff deduced certain mathematical expressions for defining the variation of heat of reaction with temperature. All these deductions are derived from the first law of thermodynamics . Consider a process,

$$A \rightarrow B$$

Let E_A and H_A be the internal energy and heat content of the system in the initial state A.

Similarly, let E_B and H_B be the internal energy and the heat content of the system in the final state B. By the first law of thermodynamics, the change in enthalpy for the given process is given by the following equation

$$\Delta H = H_B - H_A \qquad ...(21)$$

But heat content is a function of temperature and pressure. Therefore, the equation (21) can be differentiated with respect to temperature at constant pressure.

$$\left[\frac{\partial(\Delta H)}{\partial T}\right]_P = \left[\frac{\partial(\Delta_B)}{\partial T}\right]_P - \left[\frac{\partial(H_A)}{\partial T}\right]_P = (C_p)B-(C_P)A$$

$$\left[\because \left(\frac{\partial H}{\partial T}\right)_P C_P\right]$$

or
$$\left[\frac{\partial(\Delta H)}{\partial T}\right]_P = \Delta C_P \qquad ...(22)$$

where ΔC_p represents the difference in the heat capacities of initial and final state at constant pressure. In order to integrate this equation (22), we have to consider that

(i) ΔC_p is independent of temperature

or (ii) ΔC_p is dependent upon temperature.

Let us take into consideration these cases one by one.

Case I.

When ΔC_p is constant between two temperatures T1 and T_2 : If we consider C_p to be constant and integrate equation (22) between the limits H_1 at T_1 and H_2 and T_2, we get the expression

$$\int_{H_1}^{H_2} d(\Delta H) = \int_{T_1}^{T_2} \Delta C_p \, dT, \text{ or } \left[\Delta H\right]_{H_1}^{H_2} = \left[\Delta C_P\right]_{T_1}^{T_2}$$

or $$(\Delta H_2 - \Delta H1) = \Delta C_P (T_2 - T_1) \qquad ...(23)$$

This relationship is termed as Kirchhoff's equation.

Case II.

When ΔC_P varies with temperature.

It implies that equation (22) can be integrated at constant pressure

$$\int d(\Delta H) = \int \Delta C_P . \, dT$$

or $$\Delta H = \int \Delta C_P . \, dT + K \qquad ...(24)$$

where K is a constant of integration. When T = 0. then $\Delta H_o = K$. Hence equation (24) becomes as follows :

$$\Delta H = \Delta H_o + \int \Delta C_P . \, dT \qquad ...(24A)$$

where ΔH_o represents the difference in heat contents of products and reactants at the absolute zero.

But $$(C_P)_A = \propto_A + \beta_A T + \gamma_A T^2 \; ......$$

$$(C_P)_B = \propto_B + \beta_B T + \gamma_B T^2 \; ......$$

$$\therefore \quad C_P = (C_P)_B - (C_P)_A = (\propto_B + \beta_B T + \gamma_B T^2 ...) - (\propto_A + \beta_A T + \gamma_A T^2 + ...$$

$$= (\propto_B - \propto_A) + T(\beta_B - \beta_A) + T^2 (\gamma B - \gamma A) + ...$$

i.e., $$\Delta C_P = \Delta\propto + T\Delta\beta + T^2 \Delta\gamma + \qquad ...(25)$$

If we insert this value of ΔC_P in equation (24A) and carry out the integration, we get

$$\Delta H = \Delta H_o + T\,\Delta\propto + \frac{T^2}{2}\Delta\beta + \frac{T^3}{3}\Delta\gamma + ... \qquad ...(26)$$

The values of $\Delta\propto$, $\Delta\beta$, $\Delta\gamma$, ... can be calculated from heat capacity date. Equation (26) is another form of Kirchhoff's equation. This equation can be used to evaluate ΔH_o if the value of ΔH and constant $\Delta\propto$, $\Delta\beta$, $\Delta\gamma$, ... are known.

Consider the following reaction,

$$n_1A + n_2B \rightleftharpoons n_3C + n_4D$$

where n_1 and n_2 be the number of molecules of reactants A and B and n_3 and n_4 be the number of molecules of products C and D,

Now $(C_P)_A = n_1\,(\propto_1 + \beta_1 T + \gamma_1 T^2 + \delta_1\, T^3 + \ldots\ldots)$

$(C_P)_B = n_2\,(\propto_2 + \beta_2 T + \gamma_2 T^2 + \delta_2\, T^3 + \ldots\ldots)$

$(C_P)_C = n_3\,(\propto_3 + \beta_3 T + \gamma_3 T^2 + \delta_3\, T^3 + \ldots\ldots)$

$(C_P)_D = n_4\,(\propto_4 + \beta_4 T + \gamma_4 T^2 + \delta_4\, T^3 + \ldots\ldots)$

The change in heat capacity (ΔC_P) is given as follows :

$$\Delta C_P = [\{(\propto_3 + \propto_4) - (\propto_1 + \propto_2)\}] + \{(\beta_3 + \beta_4) - (\beta_1 + \beta_2)\}T$$
$$+[\{\gamma_3 + \gamma_4) - (\gamma_1 + \gamma_2)\}\ T^2 + \ldots]\ (n_3 + n_4 - n_1 - n_2)$$

or $\Delta C_P = (\propto + \beta T + \gamma T^2 + \ldots)\ \ (n_3 + n_4 - n_1 - n_3)$

On comparing it with Kirchhoff's equation, we obtain

$$\left[\partial \frac{\partial H}{\partial V}\right]_P = \Delta C_P = \propto + \beta T + \gamma T^2 + \cdots \qquad \ldots(26A)$$

On integrating this equation, we obtain

$$\Delta H = \Delta H_o + \propto T + \beta \frac{T^2}{2} + \gamma \frac{T^3}{3} + \ldots \qquad \ldots(26B)$$

where ΔH_o is the integration constant and

$\propto = \propto_3 + \propto_4 - \propto_1 - \propto_2,\ \beta = \beta_3 + \beta_4 - \beta_1 - \beta_2,$

$\gamma = \gamma_3 + \gamma_4 - \gamma_1 - \gamma_2,\ \delta = \delta_3 + \delta_4 - \delta_1 - \delta_2,$

Equation (26B) is the general expression which defines the variation of ΔH with temperature.

THE MAXIMUM FLAME TEMPERATURE

When a fuel burns, definite quantities of products are formed and a definite amount of heat is liberated. This quantity of heat may raise the temperature of the products of combustion sufficiently so as to produce a flame. This temperature is termed as the maximum flame temperature, or calorific intensity. When the experiment is carried out at constant volume the quantity of heat liberated will be ΔE and the maximum temperature is termed explosion temperature. In case the experiment is carried out at constant pressure then heat liberated is ΔH

and the maximum temperature to which the products will be raised is termed as the flame temperature. Generally this temperature is termed as the adiabatic maximum flame temperature as it is assumed that there occurs no loss of heat in the process.

Calculation of Maximum Temperature

(1) If we assume that the reaction is very fast and the total heat evolved is used to raise the temperature of the system, then the heat capacities of the species present at the completion of reaction and the heat evolved can be used for calculating the maximum temperature by using Kirchhoff's equation in the following form,

$$\Delta H_{(reaction)} = \int_{T_2}^{T_1} \sum nC_P \, dT \, \Delta H_{(heating)}$$

where it is assumed that $\Delta H_{(heating)} = -\Delta H_{(reaction)}$, as the heat evolved is taken by the products. T_1 represents the initial temperature and T_2 the maximum flame temperature, $\Sigma(nC_P)$ the sum of the heat capacities of all the species present in the system after the reaction gets completed. For constant volume process,

$$\Delta E = \int_{T_2}^{T_1} \sum nC_V \, dT$$

The direct way to know the maximum flame temperature is to carry out the experiment in the laboratory under adiabatic conditions and measure the temperature using standard techniques of temperature measurements.

The calcualted flame temperature has been found to be higher than the observed temperature. This difference has been attributed to a number of complications which are as follows :

(i) The system may not be truly adiabatic.

(ii) In combustion, we generally use excess of air or oxygen, Unreacted air or oxygen is to be heated and therefore a part of the heat would be utilised in heating these gases.

(iii) When flame temperature is to be calculated, heating of the unreacted gases should also be considered.

(iv) At high temperatures, the products might decompose and a part of the heat may be utilised to decompose the product molecules.

(2) If an exothermic reaction is assumed to carry out under adiabatic conditions, the heat evolved would be able to raise the temperature of the system.

It is possible to determine the rise in temperature by considering the reaction to take place in the following two steps, both at constant pressure.

Step I : Reactants (T_o, P) → Products (T_o, P) ΔHT_o

Step II : Products (T_o, P) → Products (T_f, P)

The heat involved in the second step may be given as follows :

$$\Delta H_2 = \int_{T_o}^{T_f} C_P \text{ (Products) } dT$$

As the overall reaction refers to the sum of the above two steps, *i.e.*,

Reactants (T_o, P) → Products (T_f, P)

the net heat change is given as follows :

$$\Delta H = \Delta H_{To} + \Delta H_2 \quad \text{...(i)}$$

As the reaction is carried out under adiabatic conditions, it is evident that $\Delta H = 0$. Hence Eq. (i) may be put as follows :

$$\Delta H_{To} + \Delta H_2 = 0.$$

or $\Delta H_2 = -\Delta H_{To}$ or

$$\int C_P \text{ (Products) } dT = -\Delta H_{To}$$

If C_Ps are regarded to be independent of temperature, then

$$C_P \text{ (Products) } (T_f - T_o) = -\Delta H_{To} \text{ or } T_f = \frac{-\Delta H_{TO}}{C_p\text{(products)}} + T_o.$$

However, the approximation that C_Ps are independent of temperature has not been correct because T_f is a very large quantity (ΔH_{To} ~ kJ and CP ~ J) of the order of thousands of kelvin.

If the reaction is carried out in a closed vessel *i.e.*, under the condition of constant volume, then ΔE and C_V replace ΔH and C_P respectively, *i.e.*,

$$T_f = \frac{(-\Delta E_{To})}{C_v \text{ (products)}} + T_o$$

EXPLOSION TEMPERATURE

When the combustion of hydrocarbons is done at constant volume under adiabatic conditions, the maximum temperature attained is termed as the explosion temperature. The explosion temperature is calculated in the same way as that of flame temperature except ΔH is replaced by ΔE. The pressure which the reacting system may acquire at the explosion temperature is termed as the maximum explosion pressure.

BOND ENTHALPIES OR ENERGIES

The bond energy of a particular type of bond in a molecule may be defined as the amount of energy required to dissociate or break one mole of that type present in the compound and separate the resulting atoms or radicals from one another. For example, the bond energy of the H – H bond in H_2 is 433 kJ mol^{-1}. It means that 433 kJ of energy is required to break or dissociate the H – H bond in one mole of hydrogen molecules. Similarly, the bond energies of I – I in I_2 and H – I in HI are 151 kJ mol^{-1} and 229 kJ mol^{-1} respectively.

In case of diatomic molecules the bond energy is a definite quantity as there is only one type of bonds present in these molecules. However, the bond energy becomes a much less definite quantity in case of polyatomic molecules. This is to be expected since in a polyatomic molecule, the energy necessary to rupture a given bond will depend to some extend on the nature of remainder molecule. For example, the energy required to rupture the O – H bond in alcohol would not be expected to the same as the energy needed to rupture same bond in an ionic compound. This is obvious from the fact that alcohols are much less acidic than organic aids. Therefore, it follows that bond energy for a particular bond in polyatomic molecules will vary from compound to compound. However, if there are similar types of bonds present in a ployatomic molecule, the bond energy is equal to the average energy required to break a particular bond in one mole of that compound. For example, C – H bond energy in CH_4 can be calculated as follows :

When methane is formed it elements carbon and hydrogen, 1648.7 kJ of heat is liberated. It means 1648.7 kJ of energy must be supplied to one mole of methane to break four identical C – H bonds, resulting the formation of atomic carbon and atomic hydrogen in the gaseous state. It means that the average C – H bond energy is 1648.7/4 or 412.175 kJ mol^{-1}. This is also the value of heat of formation of C – H bond,

From the above discussion, it is evident that the bond energy or enthalpy should be distinguished carefully from the term bond dissociation energy of enthalpy.

Bond Enthalpy or Energy

Bond enthalpy of a given bond is defined as the average of enthalpies or energy required to dissociate the said bond present in different gaseous compounds into free atoms or radicals in the gaseous state.

Bond Dissociation Enthalpy or Energy

Bond dissociation enthalpy or energy may be defined as the enthalpy or energy required to dissociate a given bond of some specific compound.

The distinction between bond enthalpy and bond dissociation enthalpy may be better understood by a simple example, say of the O – H bond. The enthalpy of dissociation of the O – H bond depends upon the nature of molecular species from which the H atom is getting separated. For example, in the water molecule

$$H_2O(g) \rightarrow H(g) + OH(g)\ ;\ \Delta H^\circ = 501.87 \text{ kJ noet}$$

$$OH(g) \rightarrow O(g) + H(g)\ ;\ \Delta H^\circ = 423.38 \text{ kJ noet}$$

The bond enthalpy of O – H is defined as the average of these values, *i.e.*,

$$\text{Bond enthalpy} = \frac{501.87 + 423.38}{2} = 462.625 \text{ kJ noet}$$

Hence the bond enthalpy given for any particular pair of atom is the average value of the dissociation enthalpies of the bond for a number of molecules in which the pair of atoms appear.

The bond dissociation energies for the stepwise dissociation of CH_4 are as follows :

$$CH_4(g) \rightarrow CH_3(g) + H(g);\ \Delta H^\circ = 427.0 \text{ kJ}$$

$$CH_3(g) \rightarrow CH_2(g) + H(g);\ \Delta H^\circ = 418.4 \text{ kJ}$$

$$CH_2(g) \rightarrow CH(g) + H(g);\ \Delta H^\circ = 460.2 \text{ kJ}$$

$$CH(g) \rightarrow C(g) + H(g);\ \Delta H^\circ = 343.1 \text{ kJ}$$

The energy for each step is different due to the fact that in each case a different dissociating fragment is involved. The C – H bond enthalpy would be one-fourth of the enthalpy required to break a mole of CH_4 and is 412 kJ.

Calculation of Bond Enthalpy or Energy

It is possible to obtain bond energies from data on heats of combustion and heats of dissociation. Let us illustrate it by calculating the bond enthalpy or energy of C – H. We know,

$$CH_4(g) \rightarrow C(g) + 4H(g) \in_{C-H} = \frac{\Delta H}{4}$$

The ΔH and hence $\in_{C-H}$ for this reaction can be obtained by adding the following equations :

$$CH_4(g) + 2O_2(g) \rightarrow CO_2(g) + 2H_2O(l) \quad \Delta H^o = -890.36 \text{ kJ}$$

$$CO_2(g) \rightarrow C \text{ (graphite)} + O_2(g) \quad \Delta H^o = 393.51 \text{ kJ}$$

$$2HO_2(l) \rightarrow 2H_2(g) + O_2(g) \quad \Delta H^o = 571.70 \text{ kJ}$$

$$2H_2(g) \rightarrow 4H(g) \quad \Delta H^o = 871.86 \text{ kJ}$$

$$C \text{ (graphite)} \rightarrow C(g) \quad \Delta H^o = 716.68 \text{ kJ}$$

$$CH_4(g) \rightarrow C(g) + 4H(g) \quad \Delta H^o = 1663.39 \text{ kJ}$$

$$\text{Thus, at } 298.15\text{o k } \in_{C-H} = \frac{1663.39}{4} = 415.85 \text{ kJ}$$

This value of the C – H bond energy does not correspond to the dissociation energy of the carbon-hydrogen bond in methane, which is 426.77 kJ mol^{-1} and refers to the equation

$$CH_4(g) \rightarrow CH_3(g) + H(g)$$

The data on bond energies are used to calculate the approximate heat of formation of a compound of known structure by adding the appropriate bond energies. Wherever direct experimental data are not available, approximate values of enthalpies of reaction can also be obtained by using bond enthalpy data.

Estimation of the Enthalpy of Formation from Bond Enthalpies

Let us calculate the heat of formation of $Se_2CI_2(g)$ from bond enthalpies by involving the following steps ; as the bond energy refers to the dissociation of Se – Se CI gas into the gaseous atoms, the enthalpy change for the formation of this gaseous molecule should be as follows :

$$2Se(g) + 2CI(g) \rightarrow Se_2CI_2(g)$$

$$\Delta H = -[\in Se - Se + 2 \in Se - CI] = -694.54 \text{ kJ mol}^{-1}.$$

However to estimate the heat of formation it becomes necessary to add two reactions in the above equation, as by definition the heat of formation refers to the elements in their standard states. Therefore, the following enthalpy changes are introduced to convert the elements from their standard states to the gaseous atoms at 298°K

$$Cl_2(g) \rightarrow 2Cl(g) \; \Delta H = 243.30 \text{ kJ mol}^{-1}$$

$$2 \text{ Se (hexagonal)} \rightarrow 2 \text{ Se(g)} \; \Delta H = 2 \times 2 \times 202.5 \text{ kJ mol}^{-1}$$

By adding these in the preceding expression, we obtain

$$2 \text{ Se (hexagonal)} + Cl_2(g) \rightarrow Se_2Cl_2(g); \; \Delta H_f = -46.22 \text{ kJ mol}^{-1}$$

Estimation of the Heat of Reaction from Bond Enthalpies or Energies

Let us illustrate this by calculating the enthalpy change for the following equation from the bond energy data :

$$C_2H_4(g) + HCl(g) \rightarrow C_2H_5Cl(g)$$

The enthalpy change for the above reaction can be calculated as follows :

ΔH = Energy required to break reactants into + Energy released to form products from the gaseous atoms gaseous atoms

$$= [4 \in_{C-H} + \in_{C=C} + \in_{H-Cl}] + [-5 \in_{C-H} - \in_{C-C} - \in_{C-Cl}]$$

$$= (\in_{C=C} + H - Cl) - (\in_{C-H} + \in_{C-C} + \in_{C-Cl})$$

On substituting the bond enthalpy values in the above equation, we get

$$\Delta H = (615.05\text{kJ} + 431.79\text{kJ}) - (413.38\text{kJ} + 347.69\text{kJ} + 328.44\text{kJ})$$

$$= -42.67\text{kJ}$$

Bond Energies and Resonance

There occurs a fair agreement between the calculated values of heat of formation obtained from bond enthalpies and any other method. However, there occurs large deviations in case of compounds having alternate single and double bonds. For example, the following reaction.

$$\underset{\text{Benzene}}{C_6H_6(g)} \rightarrow 6C(g) + 6H(g)$$

will require enthalpy of 5368.5 kJ on the basis of bond energies ($3 \in_{C=C} + 3\in_{C=C} + 6\in_{C-H}$ = 3 × 347.69 kJ + 3 × 615.05 kJ + 6 × 413.38 kJ = 5368.5 kJ), while the experimental value is 5351.1 kJ. This amounts to the fact that benzene is more stable by 166.6 kJ mol^{-1}. This is attributed to resonance, that is, in benzene there occurs no localization of single and double bonds, but the molecule is resonating from one extreme to another.

Factors: The value of the bond energy depends upon the following factors :

(a) *Length of the bond :* In general, it is found that greater the bond length between the two nuclei of the two atoms forming the bonds, the smaller is the bond energy.

Type of bond	H – H	CI – CI	I – I
Bond length (i)	0.74	1.25	2.7
Bond energy (kJ/mol)	433.1	330.5	151.2

(b) *Bond polarity :* Higher is the difference in electronegativity between the combining atoms, greater is the strength of the dipole along the bond and higher is the bond energy.

(c) *Unsaturation :* In general it is observed that with the increase in the number of covalent bonds between the atoms, there occurs an increase in the bond energy but the contribution of each additional bond goes on decreasing.

Bond	***Energy in kJ/mole***
C – C	347.3
C = C	615.0
C = C	811.7

(d) *Resonance effect :* If the bonds in the molecule do not remain fixed but the molecule possesses resonance, the bond energy per bond depends upon the amount of resonance.

(e) *Overlap of Orbitals :* The bond energy depends upon the s character which is determined by extent of overlap of orbitals of the combining atoms.

LAW OF THERMOCHEMISTRY

1. Lavoisier and Laplace's Law. This law was given by Lavoisier and Laplace in 1780 and may be stated as follows :

"The amount of heat supplied to decompose a compound into its elements is equal to the heat of formation of that compound form its elements".

In simple words, the heat of formation of compound is numerically equal to its heat of decomposition of a compound but of opposite sign. Thus, the larger the heat of formation of a compound, the greater will be its stability.

Importance : The importance of this law is that thermochemical equations can be reversed. For example, the equation

$$C(s) + O_2(g) \rightarrow CO_2(g)$$
$$\Delta H^o = -296.64 \text{ kJ}$$

may be written as

$$CO_2(g) \rightarrow C(s) + O_2(g)$$
$$\Delta H = +296.64 \text{ kJ}$$

This law applies not only to reactions involving a compound and its component elements but also to reactions of all other types. Thus, for example,

$$C_2H_5OH + 3O_2 \rightarrow 2CO_2 + 3H_2O$$
$$\Delta H^o = -1366.95 \text{ kJ}$$

Hence by Lavoisier-Laplace's law, we get

$$2CO_2 + 3H_2O \rightarrow C_2H_5OH + 3O_2$$
$$\Delta H^o = +1366.95 \text{ kJ}$$

Hess's Law of Constant Heat Summations

Most calculations involving heat changes in thermochemical equations are based on the fundamental law enunciated by Russian chemist G. H. Hess in 1840. This law can be stated in words as follows :

"The enthalpy change of a given chemical reaction is the same whether the process takes place in one or several steps".

In other words, the law is merely a corollary of the law of conservation of energy which implies that heat of reaction does not depend upon the nature of intermediate products but depends only on the initial reactants and final products and is also independent of the path or the manner by which the change is carried out.

Experimental Illustration : In order to bring about experimental illustration, we can carry out the same chemical change in two alternative ways and show that the enthalpy change remains same for both. For example, carbon dioxide may be produced by two process :

First Procedure : The burning of carbon can take place in a single step to form CO_2 with the enthalpy change which is equal to – 393.0 kJ mol^{-1}.

$$C(graphite) + O_2(g) \rightarrow CO_2(g); \Delta H = -393 \text{ kJ} \quad ...(1)$$

Second Procedure. It is also possible to carry out the above reactions in two steps. First of all carbon can be converted into carbon monoxide. Then carbon monoxide can be oxidized to carbon dioxide. Thus,

$$C(graphite) + 1/2O_2(g) \rightarrow CO(g); \Delta H = -110.5 \text{ kJ} \quad ...(2)$$

$$CO(g) + 1/2O_2(g) \rightarrow CO_2(g); \Delta H = -283.2 \text{ kJ} \quad ...(3)$$

On adding equations (2) and (3), we get

$$C(graphite) + O_2(g) \rightarrow CO_2(g); \Delta H = -393.7 \text{ kJ} \quad ...(4)$$

From equations (1) and (4), it follows that the enthalpy change in both the procedures is almost same. The small difference can be attributed to experimental error. It also follows from the above discussion that thermochemical equations can be added or subtracted as ordinary algebraic equations.

Theoretical Justification : (Hess's law as a corollary of First Law of Thermodynamics). The law also follows as a mere consequence of the First Law of Thermodynamics. Suppose a substance A can be changed to Z directly.

$$A \rightarrow Z + Q_1 \quad ...(5)$$

where Q_1 represents the heat evolved in the direct change. If the same change is carried out through intermediate stages,

$$A \rightarrow B + q_1, B \rightarrow C + q_2, C \rightarrow Z + q_3$$

The total evolution of heat = $q_1 + q_2 + q_3 = Q_2$

By Hess's Law, $Q_1 = Q_2$.

If it is not so, suppose $Q_2 > Q_1$. Then by transforming A to Z through stages and retransforming directly back to A there would occur a gain of heat = $Q_2 - Q_1$. If we repeat this process again and again, an unlimited heat will be produced in an isolated system. This goes against the First Law of Thermodynamics. Hence Q_1 must be equal to Q_2 (Fig. 1.4)

Applications of Hess's Law

The main practical use of Hess's law lies in the fact that it makes legitimate to add up or subtract thermochemical equations as algebraic equations. This gives rise to the mathematical method of calculating :

(1) Heat of Reaction,

(2) Heat of Transition,

(3) Heat of Formation, and

(4) Lattice Energy of a Crystal.

Let us describe these one by one.

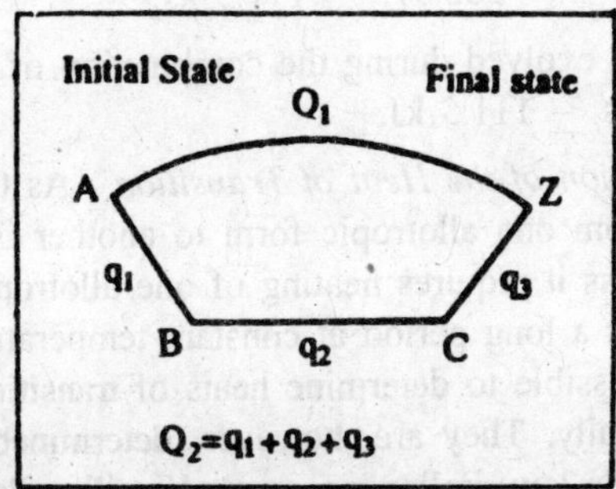

Fig. 1.4 : Theoretical Justification of Hess's law

The importance of Eq. (5) is that the amount of heat absorbed at constant volume can be identified with the change in the thermodynamic quantity.

(1) *Calculation of heats of reactions :* By Hess's law it is possible to calculate heats of many reactions which cannot be determined experimentally. For example it is very difficult to measure the heat evolved when carbon burns in oxygen to form carbon monoxide,

$$C(s) + 1/2O_2(g) \rightarrow CO(g); \Delta H = ?$$

From the Hess's law, it is evident that the heat produced in the combustion of carbon to give carbon dioxide is the same, viz., – 393.5 kJ, whether it takes place in a single step as

$$C(s) + O_2(g) \rightarrow CO(g); \Delta H = -393.5 \text{ kJ}$$

or in two steps as

$$C(s) + 1/2O_2(g) \rightarrow CO(g); \Delta H = x \text{ kJ (say)}$$
$$CO(g) + 1/2O_2(g) \rightarrow CO_2(g); \Delta H = y \text{ kJ (say)}$$

Evidently

$$x + y = -393.5 \text{ kJ}$$

The x, the heat produced in the conversion of carbon to carbon monoxide cannot be determined but y, the heat change involved in the combination of carbon monoxide and oxygen to give carbon dioxide can be measured and has been found to be – 282.0 kJ

$$x = -393.5 - y$$

$$\text{or } y = -393.5 - (-282.0) = -111.5 \text{ kJ.}$$

Hence the heat evolved during the combination of carbon to give carbon monoxide is – 111.5 kJ.

(2) *Determination of the Heat of Transition* : As transition of an element from one allotropic form to another is an extremely slow process it requires heating of one allotropic form of the element for a long period at constant temperature. Therefore, it is not possible to determine heats of transition of elements experimentally. They are, however, determined indirectly by using Hess's law. Following example will make this clear.

Suppose we intend to determine the heat of transition of sulphur (rhombic) into sulphur (monoclinic), we start with the two allotropic forms separately and measure their heats of combustion.

$$S_{(R)} + O_2(g) \rightarrow SO_2 \text{ ; } \Delta H = \text{AJ} \qquad \text{...(i)}$$

$$S_{(M)} + O_2(g) \rightarrow SO_2 \text{ ; } \Delta H = \text{BJ} \qquad \text{...(ii)}$$

On subtracting (ii) from (i), we get

$$S_{(R)} - S_{(M)} \rightarrow 0 \text{ ; } \Delta H = (A - B) \text{ J}$$

$$S_{(R)} - S_{(M)} \text{ ; } \Delta H = (A - B) \text{ J} \qquad \text{...(iii)}$$

Equation (iii) expresses the thermochemical equation for the transition of $S_{(R)}$ to $S_{(M)}$ Hence heat of transition $S_{(R)} \rightarrow S_{(M)}$ is (A – B) J.

In this particular case, $\Delta H = 2.5$ kJ.

(3) *Calculation of Heats of Formation* : The heat of formation of those compounds which cannot be determined experimentally can be calculated by using Hess's law. This can be understood by considering the following example:

The enthalpy of formation of methane from graphite and hydrogen cannot be found directly as no chemical reaction occurs between hydrogen and graphite.

(i) $C(\text{graphite}) + 2H_2(g) \rightarrow CH_4(g)$

However, ΔH for the above reaction may be calculated by the help of the following reactions :

(ii) C(graphite) + O_2(g) → CO_2(g); ΔH_a = – 393.5 kJ

(iii) $2H_2$(g) + O_2(g) → $2H_2O$(l); ΔH_b = 2(–285.9 kJ) = – 571.8 kJ

(iv) CH_4(g) + $2O_2$(g) → CO_2(g) + $2H_2O$(*l*); ΔH_c = – 890.3 kJ

From the above reactions, the enthalpy change for reaction (i) may be obtained by adding equations (ii) and (iii)

(ii) C(graphite) + O_2(g) → $3CO_2$(g); ΔH_a = – 393.5 kJ

(iii) $2H_2$(g) + O_2(g) → $2H_2O$(l); ΔH_b = – 571.8 kJ

C(graphite) + $2O_2$(g) + $2H_2$(g) → CO_2(g) + $2H_2O$(l);

$\Delta H_a + \Delta H_a$ = – 965.3 kJ

On subtracting equation (iv) from the above equation, we obtain

C(graphite) + $2O_2$(g) → CH_4(g);

ΔH = – 965.3kJ + 890.3kJ = – 75.0 kJ

(4) *Lattice Energy of a crystal (Born-Haber Cycle)* : Hess's law can be used to determine lattice energy which may be defined as the amount of heat released when the requisite amounts of ions in the gaseous state combine to yield 1 mole of crystal lattice. An example is the determination of $DH_{Lattice}$ of sodium chloride crystal which corresponds to the reaction.

$$Na^+(g) + CI^-(g) \rightarrow NaCI\ (s).$$

It is possible to determine the numerical value of $\Delta H_{Lattice}$ for the above reaction directly by using Born-Haber's cycle. The following is the sequence of steps for the formation of NaCI crystals from Na(s) and CI_2(g)

(i) Vaporisation of Na(s)	Na(s) → Na(g)	ΔH_1
(ii) Ionisation of Na(g)	Na(g) → Na^+(g)+e	ΔH_2
(iii) Dissociation of chlorine	1/2 CI_2(g) → CI(g)	ΔH_3
(iv) Formation of CI^-	CI(g) + e → CI^-(g)	ΔH_4
(v) Condensation of	Na^+(g) and CI^-(g)	Na+(g)
	+ CI–(g) → NaCI(s)	ΔH_5
Net change	Na(s)+1/2 CI_2(g) NaCI(s)	ΔH_6

On applying the Hess's law to the above sequence of steps, we get

$$\Delta H_6 = \Delta H_1 + \Delta H_2 + \Delta H_3 + \Delta H_4 + \Delta H_5$$

Except ΔH_5, it is possible to determine experimentally all of the changes of enthalpy, *i.e.*, ΔH_1, ΔH_2, ΔH_3, ΔH_4, and ΔH_6. Hence ΔH_5 can be determined from the above relation.

SOLVED EXAMPLES

Example 1:

Calculate ΔH^o for the reaction at 298 K

$$Ba^{2+}(aq) + SO_4^{2-} \rightarrow BaSO_4(s)$$

Given that

$$\Delta H_f^o\,[Ba^{2+}(aq)] = -\,538.36\ kJ$$

$$\Delta H_f^o\,[SO_4^{2-}(aq)] = -\,907.51\ kJ$$

$$\Delta H_f^o\,[BaSO_4(s)] = -\,1465.24\ kJ$$

Solution:

For the reaction,

$$Ba^{2+}(aq) + SO_4^{2-} \rightarrow BaSO_4(s)$$

We have $\Delta H^o = \Delta H_f^o\,[BaSO_4(s)] - \Delta H_f^o\,[Ba^{2+}(aq)] - \Delta H_f^o\,[SO_4^{2-}(aq)]$

On substituting the values, we get

$$\Delta H^o = -\,1465.24\ kJ - (-\,538.36\ kJ - 907.51\ kJ) = -\,19.37\ kJ$$

Example 2:

Calculate the bond enthalpy for a C – O bond in methanol from the following data :

(i) $C(s)+2H_2(g)+1/2O_2 \rightarrow CH_3OH(g);\quad \Delta H^o = -\,200\ kJ\ mol^{-1}$

(ii) $C(s) \rightarrow C(g);\quad \Delta H^o = -\,716.8\ kJ\ mol^{-1}$

(iii) $2H_2(g) \rightarrow 4H(g);\quad \Delta H^o = -872.0\ kJ\ mol^{-1}$

(iv) $1/2O2(g) \rightarrow O(g);\quad \Delta H^o = -249.0\ kJ\ mol^{-1}$

The bond enthalpy for C – H bond is 403 kJ mol^{-1} and for O–H bond it is 463.6 kJ mol^{-1}.

Solution:

The enthalpy change for the dissociation of CH_3OH is as follows :

$CH_3OH(g) \rightarrow C(g) + 4H(g) + O(g)$; $\Delta H^o = ?$

On adding (ii), (iii) and (iv), we get

$C(s) + 2H_2(g) + 1/2O_2(g) \rightarrow C(g) + 4H(g) + O(g)$,

$\Delta H^o = 1837.8$ kJ

On subtracting equation (i) from the above equation, we get

$CH_3OH(g) \rightarrow C(g) + 4H(g) + O(g)$;

$\Delta H^o = 2037.8 kJ\ mol^{-1}$

As methanol has three C – H, one C – O bond and one O – H bond, it means that

$2037.8\ kJ = 3\ (\Delta H^o_{C-H}) + (\Delta H^o_{O\ H}) + 1\ (\Delta H^o_{C-O})$

$= 3\ (413) + (463.6) + ((\Delta H^o_{C-O})$

or $(\Delta H^o_{C-O}) = 335.2\ kJ\ mol^{-1}$

Example 3:

The enthalpy of neutralisation of ammonium hydroxide by hydrochloric acid is 51.46 kJ mol⁻¹. Calculate the enthalpy of ionisation of ammonium hydroxide.

Solution:

(i) $NH_2OH(aq) + H^+(aq) \rightarrow NH_4^+(aq) + H_2O(l)$; $\Delta H^o = -51.46 kJ\ mol^{-1}$

(ii) $H^+(aq) + OH^-(aq) \rightarrow H_2O(l)$; $\Delta H^o = -57.32 kJ\ mol^{-1}$

(iii) $NH_2OH(aq) \rightarrow NH_4^+(aq) + OH–(aq)$; $\Delta H^o = \Delta H^o_{Ionisation}$

On adding (ii) and (iii), we get

$NH_2OH(aq) + H^+(aq) \rightarrow NH_4^+(aq) + H_2O(l)$

or $\Delta H^o = -\ 57.32 + \Delta H^o_{Ionisation}$

From Eq. (i), we have

$\Delta H^o = -\ 51.46\ kJ\ mol^{-1}$

Hence $(-\ 57.32 + \Delta H^o_{Ionisation})\ kJ\ mol^{-1} = -\ 51.46\ kJ\ mol^{-1}$

$\Delta H^o_{Ionisation} = +\ 5.86\ kJ\ mol^{-1}$.

Example 4:

The bond enthalpy of $H_2(g)$ is 436 kJ mol^{-1} and that of $N_2(g)$ is 941.3 kJ mol^{-1}. Calculate the average bond enthalpy of an N – H bond in ammonia ΔH_f^o (NH_3) = – 46.0 kJ mol^{-1}.

Solution:

∵ (i) $N_2(g) \rightarrow 2N(g)$; $\Delta H^o = 941.3$ kJ mol^{-1}

(ii) $H_2(g) \rightarrow 2H(g)$; $\Delta H^o = 436.0$ kJ mol^{-1}

(iii) $1/2N_2(g) + 3/2H_2(g) \rightarrow NH_3(g)$; $\Delta H^o = -46.0$kJ mol^{-1}

On multiplying equation (i) by 1/2 and equation (ii) by 3/2 and adding, we get

(iv) $1/2N_2(g) + 3/2H_2(g) \rightarrow N(g)+3H(g)$; $\Delta H^o = 1124.6$kJ mol^{-1}

On subtracting equation (iii) from (iv), we get

(v) $NH_3(g) \rightarrow N(g) + 3H(g)$, $\Delta H^o = 1170.6$ kJ mol^{-1}

As there are three N – H bonds, in NH_3, the average bond enthalpy is obtained by dividing the value ΔH° from (v) by 3. Hence

$$\Delta H_{N-H} = 1170.6/3 = 390.2 \text{ kJ mol}^{-1}$$

Example 5:

0.50 f of benzoic acid was subjected to combustion in a bomb calorimeter at 15°C when the temperature of the calorimeter system (including water) was found to rise by 0.55°C. Calculate the heat of combustion of benzoic acid : (i) at constant volume, (ii) at constant pressure. The thermal capacity of the calorimeter system including water was found to be 23.85 kJ.

Solution:

(i) Heat of Combustion at Constant Volume (ΔE)

Weight of benzoic acid, m = 0.50 g

Rise in temperature, θ = 0.55°C

Molecular weight of benzoic acid, M = 122

Thermal capacity of the calorimeter system, z = 23.85 kJ

$\therefore$ Heat of combustion at constant volume, ΔE $\Big] = z \times \theta \times M/m$

$= -23.85 \times 0.55 \times \frac{122}{0.50} = -3200.7 \text{ kJ} = -3200.7 \text{ kJ}$

(ii) Heat of Combustion at constant Pressure (ΔH)

The combustion of benzoic acid at 25°C involves the following reaction :

$C_6H_5COOH\ (s) + 7.5O_2(g) \rightarrow 7CO_2(g) + 3H_2O(l)$

We known that

$\Delta H = \Delta E + \Delta nRT$

In the present case :

$\Delta E = -3200.7$ kJ $\Delta n = 7 - 7.5 = -2.5$ $R = 8.314 \times 10^{-3}$ kJ

$T = 273 + 25 = 298°K$

$\therefore \Delta H = -3200.7 + (-0.5) \times 8.314 \times 10^{-3} \times 298 = -3201.9$ kJ

$\therefore$ Heat of Combustion at constant pressure,

$\Delta H = -3201.9$ kJ

Example 6:

The heats of formation of carbon dioxide and water are 395 kJ and 285 kJ respectively, and the heat of combustion of acetic acid is 869 kJ (all exothermic). Calculate the heat of formation of acetic acid.

Solution:

The date given are expressed in the form of three thermochemical equations :

$C + O_2 = CO_2$ $\qquad \Delta H_1 = -395$ kJ (1)

$H_2 + 1/2O_2 = H_2O$ $\qquad \Delta H_2 = -285$ kJ (2)

$CH_3COOH + 2O_2 = 2CO_2 + 2H_2O$ $\quad \Delta H_3 = -869$ kJ (3)

The equation corresponding to the formation of acetic acid from its elements is

$2C + 2H_2 + O_2 = CH_3COOH \qquad \Delta H_f = ?$

This equation may be obtained from the three thermochemical equations on multiplying equations (1) and (2) by two and adding, finally subtracting equation (3) :

$$2 \times (1) \qquad 2C + 2O_2 = 2CO_2$$

$$2 \times (2) \qquad 2H_2 + O_2 = 2H_2O$$

Adding $\quad 2C + 3O_2 + 2H_2 = 2CO_2 + 2H_2O$

Subtract (3) $\quad CH_3COOH + 2O_2 = 2CO_2 + 2H_2O$

Giving, after re-arranging $\quad 2C + O_2 + 2H_2 = CH_3COOH$

$2 \times (1) \qquad 2 \times \Delta H_1 = -790$ kJ

$2 \times (2) \qquad 2 \times \Delta H_2 = -570$ kJ

Adding $\quad \Delta H = -1360$ kJ

Subtract (3) $\quad \Delta H_3 = -896$ kJ

Giving $\quad \Delta H_f = -2360 - (-869) = -491$ kJ

Example 7:

Heats of neutralization (ΔH) of NH_4OH and HF are – 51.5 and 68.6 kJ respectively. Calculate their heats of dissociation.

Solution:

We have

(i) $HCl(aq) + NaOH(aq) \rightarrow NaCl(aq) + H_2O; \quad \Delta H = -57.3$ kJ

(ii) $HCl(aq) + NH_4OH(aq) \rightarrow NH_4Cl(aq) + H_2O; \Delta H = -51.5$ kJ
(Weak base)

$\therefore$ The heat of dissociation of NH4OH,

$$\Delta H = 51.5 - (57.3) = 5.8 \text{ kJ}.$$

Similarly we have

$HF(aq) + NaOH(aq) \rightarrow NaF + H2O;$

$$\Delta H = -68.6 \text{ kJ}$$

$\therefore$ The heat of dissociation of HF,

$$\Delta H = -68.6 - (57.3) = -11.3 \text{ kJ}.$$

Example 8:

Calculate the heat of reaction at 25°C for the reaction :

$$C_2H_4(g) + H_2(g) \rightarrow C_3H_6(g)$$

given that the heats of combustion of ethylene, hydrogen and ethane at 25oC are 337.0, 68.4 and 373.0 kcal respectively.

Solution:

The thermochemical equations for the combustion of ethylene, hydrogen and ethane may be written as

(i) $C_2H_4 + 3O_2 \rightarrow 2CO_2 + 2H_2O$ $\Delta H = -337$ kcal

(ii) $H_2 + 1/2O_2 \rightarrow H_2O$ $\Delta H = -68.4$ kcal

(iii) $C_2H_6 + 31/2O_2 \rightarrow 3CO_2 + 3H_2O$ $\Delta H = -337$ kcal

$$(i) + (ii) - (iii) \Rightarrow C_2H_4 + H_2 - C_2H_6$$

Heat of reaction = – 337 – 68.4 + 373 = – 32.3 kcal.

Example 9:

If E_{C-C} is 34 kJ mole^{-1} and $E_{C\ H}$ is 415 kJ mole^{-1}, calculate the heat of formation of propane. The heats of atomization of carbon and hydrogen are 716 kJ mole^{-1} and 433 kJ mole^{-1} respectively.

Solution:

The heat of formation is the sum of the heats of atomization and bond energies. For propane, the heats of atomization as

$$3C_s = 3C_g; \quad \Delta H = 3 \times 716 = 2148 \text{ kJ}$$

$$4H_{2,g} = 8H_g; \Delta H = 4 \times 433 = 1732 \text{ kJ}$$

The bond energies are

$$2E_{C-C} = 2 \times -344 = -688 \text{ kJ}$$

$$2E_{C-H} = 8 \times -415 = -3320 \text{ kJ}$$

Adding,

$$3C + 4H_2 = C_3H_8; \quad \Delta H_f = 2148 + 1732 - 688 - 3320$$

$$= -128 \text{ kJ mole}^{-1}$$

Example 10:

The heats of formation of PCI_3 and PH_3 are c 306 kJ mole ^{1} and + 8 kJ mole ^{1} respectively, and the heats of atomization of phosphorus, chlorine and hydrogen are given by :

$$P_s = P_g \qquad \Delta H = 314 \text{ kJ mole}^{-1}$$

$Cl_{2.g} = 2Cl_g$ $\Delta H = 242$ kJ mole^{-1}

$H_{2.g} = 2H_g$ $\Delta H = 433$ kJ mole^{-1}

Calculate E_{P-Cl} *and* E_{P-H}

Solution:

For phosphorus trichloride, thermochemical equations are :

$P_s = P_g$ $\Delta H_1 = 314$ kJ

$3/2\ Cl_{2.g} = 3Cl_g$ $\Delta H_2 = 363$ kJ

$P_g + 3Cl_g = P_{Cl3} = 2H_g$ $\Delta H_3 = ?$

Also,

$P_s + 3/2\ Cl_{2.g} = PCl_3$ $\Delta H_f = -306$ kJ

Equating the two routes leading to the formation of PCl_3,

$$\Delta H_f = \Delta H_1 + \Delta H_2 + \Delta H_3$$

$$-306 = 314 + 363 + \Delta H_3$$

$$\Delta H_3 = -983 \text{ kJ.}$$

This is energy released on the formation of three P – Cl bonds, hence,

$$E_{P-Cl} = \frac{983}{3} = 328 \text{ kJ mole}^{-1}$$

$P_s = P_g$ $\Delta H_1 = 314$ kJ

$3/2\ H_{2,g} = 3H_g$ $\Delta H_2 = 649$ kJ

$P_2 + 3H_g = PH_3$ $\Delta H_3 = ?$

Also,

$P_s + 3/2\ H_{2.g} = PH_3$ $\Delta H_f = 8$ kJ

From which

$$\Delta H_3 = -958 \text{ kJ}$$

$$E_{P-H} = \frac{958}{3} = 319.33 \text{ kJ}$$

Example 11:

100g of anhydrous $CuSO_4$, when dissolved in excess of water produced 42 kJ of heat. The same amount of $CuSO_4$. $5H_2O$ on dissolving in large excess of water absorbed 4.60 kJ. What is the heat of hydration $CuSO_4$?

Solution:

$$100g \text{ of } CuSO_4 = \frac{100}{159.5} \text{ mole}$$

$$\text{Heat of solution of } CuSO_4 \text{ per mole} = \frac{42 \times 159.5}{100} = -66.78 \text{ kJ}$$

$$100 \text{ g of } CuSO_4.\ 5H_2O = \frac{100}{249.5} \text{ mole}$$

$$\text{Heat of solution of } CuSO_4.\ 5H_2O \text{ per mole} = 4.6 \times \frac{249.5}{100} = 11.48 \text{ kJ}$$

Thus,

(i) $CuSO_4.\ 5H_2O + aq \rightarrow CuSO_4(aq) + 66.78$ kJ

(ii) $CuSO_4.\ 5H_2O + aq \rightarrow CuSO_4(aq) - 11.48$ kJ

On subtracting (i) from (ii), we get

$$CuSO_4 + 5H_2O \rightarrow CuSO_4.\ 5H_2O + 78.26 \text{ kJ}$$

$\because$ Enthalpy of hydration of $CuSO_4$, $\Delta H = -78.26$ kJ.

Example 12:

Heat of combustion of glucose at constant pressure at 17oC was found to be –651,000 cal. Calculate the heat of combustion of glucose at constant volume considering water to be in the gaseous state.

Solution:

$$\underset{\text{negligible volume}}{C_6H_{12}O_6(s)} + \underset{6 \text{ mole}}{6O_2(g)} \rightarrow \underset{6 \text{ mole}}{6CO_2(g)} + \underset{6 \text{ mole}}{H_2O(g)}$$

$$\Delta n = 12 - 6 = 6$$

$$\Delta H = \Delta E + \Delta nRT$$

$$\Delta E = \Delta H - \Delta nRT = -651000 - 6 \times 2 \times (273 + 17)$$

$$= 651000 - 3480 = -654480 \text{ cal}$$

Hence heat of combustion of glucose at constant volume and at 17°C is – 654,480 cal.

Example 13:

The combustion of methane takes place according to the following equation :

$$CH_4(g) + 2O_2(g) \rightarrow CO_2(g) + 2H_2O(l) \;; \quad \Delta H = -890.3 \text{ kJ}$$

(a) How many grams of methane would be required to produce 89.03 kJ of heat of combustion ?

(b) How many grams of carbon dioxide would be formed when 89.03 kJ of heat is evolved ?

Solution:

(a) Molecular weight of $CH_4 = 1 \times 12 + 4 \times 1$

$= 12 + 4 = 16 \quad [\because C = 12, H = 1]$

$\because$ 890.3 kJ of energy is liberated by the combustion of 1 mole of 16 g of methane.

$\therefore$ 890.3 kJ of energy is liberated by the combustion of

$$16 \times \frac{89.03}{890.3} = 1.6 \text{ g of methane.}$$

$= 1.6$ g of methane.

(b) Molecular weight of $CO_2 = 1 \times 12 + 2 \times 16$

$= 12 + 32 = 44$

$[\because C = 12, O = 16]$

$\because$ 890.3 kJ of energy is liberated to produce 1 mole or 44 g of CO_2

$\therefore$ 890.3 kJ of energy is liberated to produce

$$44 \times \frac{89.03}{890.3} \text{ or } 4.4 \text{ g of } CO_2$$

$= 4.4$g of carbon dioxide.

Example 14:

Calculate the heat of formation of acetic acid if heat of combustion is 869.0 kJ mol^{-1}. The heats of formation of $CO_2(g)$ and $H_2O(l)$ are – 395.0 kJ mol^{-1} and – 285.0 kJ mol^{-1} respectively.

Solution:

The data given are expressed in the form of three the mochemical equations.

(1) $C(s) + O_2(g) \rightarrow CO_2(g)$; $\quad \Delta H_1 = -395$ kJ

(2) $H_2(g) + 1/2O_2(g) \rightarrow H_2O(l)$; $\quad \Delta H_2 = -285$ kJ

(3) $CH_3COOH(l) + 2O_2(g) \rightarrow 2CO_2(g) + 2H_2O(l)$; $\quad \Delta H_3 = -869$ kJ

The equation corresponding to the formation of acetic acid from its elements is

$$C(s) + 2H_2(g) + O_2(g) \rightarrow CH_3COOH(l)\ ;\quad \Delta H = ?$$

This equation may be obtained from the three thermochemical equations on multiplying equations (1) and (2) by two and adding and finally subtracting equation (3);

$$2C(s) + 2O_2(g) \rightarrow 2CO_2(g)$$

$$2H_2(g) + O_2(g) \rightarrow 2H_2O(l)$$

Adding $\quad 2C(s) + 3O_2(g) + 2H_2(g) \rightarrow 2CO_2(g) + 2H_2O(l)$

Subtract Eq. (3) $CH_3COOH(l) + 2O_2(g) \rightarrow 2CO_2(g) + 2H_2O(l)$

$$2C(s) + O_2(g) + 2H_2(g) \rightarrow CH_3COOH(l)$$

Rearranging

$$2 \times (1) 2 \times \Delta H_1 = -790 \text{ kJ}$$

$$2 \times (2) 2 \times \Delta H_2 = -570 \text{ kJ}$$

Adding $\quad \Delta H' = -1360$ kJ

Subtract Eq. (3) $\quad \Delta H_3 = -869$ kJ

Giving $\quad \Delta H = -1360 - (869) = -$ kJ.

Example 15:

The enthalpy of evaporation of water at 373 K is 40.67 kJ mol $^{-1}$. *What will be the enthalpy of evaporation at 353 K and 393 K if average molar heats at constant pressure in this range for water in liquid and vapour states are 75.312 and 33.89 JK* $^{-1}$ *mol* $^{-1}$ *respectively ?*

Solution:

$$H_2O(l) \rightleftharpoons H_2O(g)$$

Enthalpy change at 373 K is $\Delta H_1 = 40.67$ kJ mol^{-1}

$$= 40670 \text{ J mol}^{-1}$$

$$\Delta H_2 - \Delta H_1 = \Delta C_p(T_2 - T_1)$$

$$\Delta H_2 = \Delta H_1 + \Delta C_1(T_2 - T_1)$$

$$\Delta C_p = \Delta C_{p,\ H2O(g)} - \Delta C_{p,\ H2O(l)}$$

$$= 33.89 = 75.312 = -41.422.$$

Enthalpy of evaporation at 353 K, ΔH

$= 40670 - 41.522(353 - 373) = 40670 + 41.421 \times 20$

$= 41498.44$ J.

Similarly enthalpy of evaporation at 393 K

$= 40670 - 41.422(393 - 373) = 39841.56$ J.

Example 16:

The heat of formation of one mole of HI from hydrogen and iodine vapour at 25°C is 8000 cal.

$$1/2H_2(g) + 1/2I_2(g) \rightarrow HI(g),\ \Delta H = -8000 \text{ cal.}$$

Calculate the heat of formation at 10°C and also the total change in the heat capacity at constant pressure. It is given that the molar heat capacities of hydrogen, iodine and HI vapours are given by the equations :

$C_p = 6.5 + 0.0017T$	*for*	*Hydrogen(g)*
$C_p = 6.5 + 0.0031T$	*for*	*Iodine(g)*
$C_p = 6.5 + 0.0016T$	*for*	*HI(g)*

where T is the absolute temperature.

Solution:

$$\Delta C_p = C_{p(HI)} - 1/2\ [C_{p(HI)} + C_{P(I2)}]$$

$$= 6.5\ \ 0.0016T - 1/2\ [6.5 + 0.0017T + 6.5 + 0.0038T]$$

$$= -0.00115T \text{ cal. deg}^{-1}.$$

Here $\Delta H_{298} = \Delta H_{283} + \int_{283}^{298} \Delta C_p \, dT$

where $T_1 = 273 + 10 = 283$ K, $T_2 = 273 + 25 = 298$ K

Also, ΔH_{298} K $= -8000$ cal.

$$-8000 = \Delta H_{283} + \int_{283}^{298} (-0.00115)\ T\,dT$$

or $\Delta H_{283} = -8000 + \frac{0.00115}{2}[T_2]_{283}^{298} = -8000 + \frac{0.00115}{2}\left[(298)^2 - (283)^2\right]$

$= -7995$ cal (nearly).

Example 17:

How much heat is required to raise the temperature of 1 mole of oxygen from 303 K to 1300 K at constant pressure?

$$C_p = 6.095 + 3.253 \times 10^{3} T - 1.017 \times 10^{6} T^2.$$

Solution:

$$H_2 - H_1 = \int_{T_1}^{T_2} C_p \, dT$$

$$T_1 = 300 \text{ K}, T_2 = 100\text{K}.$$

$$H_{1300} - H_{300} = \int_{300}^{1300} \left(6.095 + 3.253 \times 10^{-3} T - 1.07 \times 10^{-6} T^2\right) dT$$

[Substituting the value of C_p]

$$H_{1300} - H_{300} = 6.095\,(1300 - 300)$$

$$+ \frac{3.253 \times 10^{-3}}{2} \times \left[(1300)^2 - (300)^2\right]$$

$$- 1/3 \times 1.017 \times 10{-}6\,[(1300)^3 - (300)^3]$$

$$= 6095 + 2602.4 - 735.63 = 7961.77 \text{ cal/mole.}$$

Example 18:

Calculate ΔH at 25°C for the reaction

$$H_2O(g) \rightarrow H_2(g) + 1/2O_2(g)$$

given that ΔH at 18°C is 241.750 kJ mol^{-1} and molar-Heat capacities expressed as Jk^{-1} $mole^{-1}$ are $C_p(H_2) = 28.83$, $C_p(O) = 29.12$ $C_p(H_2O) = 33.56$

Solution:

ΔC_p = Heat capacities of products – Heats capacities of reactants $= [28.83 + 1/2 \times 29.12] - 33.56 = 43.39 - 33.56 = 9.83 Jk^{-1} mole^{-1}$

$$T_2 - T_1 = (273 + 25) - (273 + 18) = 7 \text{ k}$$

$$\Delta H_2 + \Delta H_1 = \Delta C_p\,(T_2 - T_1)$$

or $$\Delta H_2 = \Delta H_1 + \Delta C_p\,(T_2 - T_1) = 241.750 + 9.83 \times 7$$

$$= 310.56 \text{ kJ mol}^{-1}.$$

Example 19:

A student made the following erroneous statement in a laboratory report on bomb calorimetry.

$$\Delta H = \Delta U + p\Delta V$$

Since the bomb calorimetry process is a constant volume one, $\Delta V = 0$ and $\Delta H = \Delta U$. Explain why this argument is incorrect

Solution:

Since, $H = U + pV$

we will have $dH = dU + pdV + Vdp$.

For a constant volume process, the above expression is reduced to

$$dH = dU + V\,dp$$

While writing the expression

$$\Delta H = \Delta U + p\,\Delta V$$

the student has not written the expression V dp which makes a significant contribution when the system involves the gaseous phase. Hence, his conclusion $\Delta H = \Delta U$ is incorrect.

Example 20:

H_2 gas is mixed with air at 25°C under a pressure of 1 atmosphere and exploded in a closed vessel. The heat of the reaction

$$H_2(g) + 1/2\ O_2(g) \rightarrow H_2O(g)$$

at constant volume, $\Delta E_{298} = -240.60$ kJ mol^{-1} and C_p values for H_2O vapour and N_2 in the temperature range 298 K and 3,200 K are 39.06 JK^{-1} mol^{-1} and 26.40 JK^{-1} mol^{-1} respectively. Calculate the explosion temperature under adiabatic conditions.

Solution:

For a constant volume process under adiabatic conditions,

$$\Delta E = \Delta E_{heating} + \Delta E_{298} = 0$$

$$\therefore \quad \Delta H_{heating} = -\Delta E_{298}$$

$$= -\int_{298}^{T_f} n \sum C_V\, dT = -240.60 \text{ kJ mol}^{-1} \qquad \text{...(i)}$$

As 2 moles of unreacted N_2 are associated with 1/2 mole of O_2, we have

$$\Sigma nC_V = C_V(H_2O, g) + 2CV\ (N_2, g)$$

$$= (39.06 + 2 \times 26.40) \text{ Jk}^{-1} \text{ mol}^{-1}$$

$$= 91.86 \text{ Jk}^{-1} \text{ mol}^{-1}$$

Hence, from Eq. (i) on integrating, we get

$$(91.86\ Jk^{-1}\ mol^{-1})\ (T_f - 298) = 240{,}600\ J\ mol^{-1}$$

$$T_f - 298 = \frac{240600\ jmol^{-1}}{91.86\ J\ k^{-1}\ mol^{-1}} = 2619\ k,$$

$$T_f = (2619 + 298)\ k = 2917\ k.$$

Example 21:

Estimate the adiabatic flame temperature of a spirit lamp, provided the following data

$$C_2H_5OH(l) + 3O_2(g) \rightarrow 2CO_2(g) + 3H_2O(g)$$

$$\Delta H_{298}\ k = 1370\ kJ\ mol^{-1}.$$

$C_{p.m}$ *for (i)* $CO_2(g) = 36.43\ J\ k^{-1}\ mol_{-1}$,

(ii) $H_2O(g) = 33.70\ J\ k^{-1}\ mol^{-}{}_{1}$ *and*

(iii) $N_2(g) = 294.4\ J\ k^{-1}\ mol_{-1}$.

If the above reaction is carried out in air consisting of 80 % N_2 *and 20% of* O_2 *by volume, what would be the final temperature ?*

Solution:

The expression of adiabatic flame temperature is

$$T_f = \frac{-\Delta H}{C_p(\text{products})} = T_o$$

$$= \frac{\left(1370 \times 10^3\ J\ mol^{-1}\right)}{(2 \times 36.43 + 3 \times 33.70\ )J\ k^{-1}\ mol^{-1}} + 298k$$

$$= 7875\ k + 298\ k = 8173\ k$$

If the reaction is carried out in air, the consumption of 3 mol of O_2 would leave 12 mol of N_2 from air. Hence,

$$T_f = \frac{\left(1370 \times 10^3\ J\ mol^{-1}\right)}{(2 \times 36.43 + 3 \times 33.70 + 12 \times 29.4\ )J\ k^{-1}\ mol^{-1}} + 298k$$

$$= 2600\ k + 298\ k = 2898\ k$$

Example 22:

Use Born-Haber cycle to determine the lattice energy of KCI(s) at 298.15 K. The following data are given at the same temperature.

Enthalpy of formation of KCl	$= -435.9$ *k J mol* ^{1}
Ionization energy of K	$= 418.9$ *k J mol* ^{1}
Electron affinity of Cl	$= 348.6$ *k J mol* ^{1}
Dissociation enthalpy of Cl_2	$= 242.0$ *k J mol* ^{1}
Sublimation enthalpy of K	$= 89.8$ *k J mol* $^{-1}$

Solution:

We are given that

(1) $K(s) + 1/2\ Cl_2(g) \rightarrow KCl(s)$ $\Delta H_1 = 435.9$ kJ mol^{-1}

(2) $K(g) \rightarrow K^+(g) + e$ $\Delta H_2 = 418.9$ kJ mol^{-1}

(3) $Cl(g) + e \rightarrow Cl^-(g)$ $\Delta H_3 = -348.6$ kJ mol^{-1}

(4) $Cl_2(g) \rightarrow 2Cl(g)$ $\Delta H_4 = 242.0$ kJ mol^{-1}

(5) $K(s) \rightarrow K(g)$ $\Delta H_5 = 89.8$ kJ mol^{-1}

We have to find ΔH of the following equation

(6) $K^+(g) + Cl^-(g) \rightarrow KCl(s)$

Equation (6) can be obtained by the following manipulations.

Eq. (1) – Eq. (5) – Eq. (2) (2) – 1/2 Eq. (4) – Eq. (3)

Accordingly, we have

$$\Delta H_6 = \Delta H_1 - \Delta H_5 - \Delta H_2 - \Delta H_1 .2\ \Delta H_4 - \Delta H_3$$

$$\Delta H6 = (-435.9 - 89.8 - 418.9 + 1/2 \times 242.0 + 348.6) \text{ kJ mol}^{-1}$$

$$= -475.0 \text{ kJ mol}^{-1}.$$

Example 23:

For the hypothetical reaction

$$2\,A\,(g) \rightarrow A_2(g)$$

$\Delta C_p / J\,k^{1}\,mol^{1} = 1.00 + (2.00 \times 10{-}3/k)\,T$ *and* $\Delta H^o{}_{298}k = -5.00$ *kJ* mol^{-1}. *Estimate the temperature at which* $\Delta H^o = 0$ *for this reaction at constant pressure.*

Solution:

We have $\int_{\Delta H^o}^{0} d(\Delta H) = \int_{298k}^{T} (\Delta C_p)\, dT$

$$= \int_{298\,k}^{T} \left[1.00 + \left(\frac{2.00 \times 10^{-3}}{k} \right) T \right] J\ k^{-1}\ mol^{-1}\ dT$$

$$0 - \Delta H^{o} = (1.00\ J\ k^{-1}\ mol^{-1})\ (T - 298\ k)$$

$$+ \left(\frac{2.00 \times 10^{-3}}{2k} J\ k^{-1}\ mol^{-1} \right) (T^2 - 298\ k^2)$$

$$5000 = (1.00\ k^{-1})\ (T - 298\ k) + (10^{-3}\ k^{-2})\ (T^2 - 298k^2) = 0$$

or $$10^{-3}\left(\frac{T}{K}\right)^2 + \left(\frac{T}{K}\right) - 298^2 \times 10{-}3 - 298 - 5000 = 0$$

or $$10^{-3}\left(\frac{T}{K}\right)^2 + \left(\frac{T}{K}\right) - 5209 = 0$$

Solving for T/K, we get

$$\left(\frac{T}{K}\right) = \frac{-1 + \left(1^2 + 4 \times 10^{-3} \times 5209\right)^{1/2}}{2 \times 10^{-3}}$$

$$T = \frac{-1 + 4.67}{2 \times 10^{-3}} k = 1836\ k$$

Example 24:

What are the reasons for the 'calculated maximum adiabatic flame temperature' not being exactly equal to theexperimental value ? What correction would you suggest ?

Solution:

The maximum adiabatic flame temperature is calculated using the expression

$$T_f = \frac{-\Delta H}{C_p(\text{products})} + T_0$$

In derving the above expression, it is assumed to C_ps are independent of temperature. However, this is not correct since T_f is a very large quantity (as ΔH is expressed in kJ whereas C_ps are in joule) of the order of thousands of kelvin. Besides this, the above value will be obtained if the combustion process is carried out reversibly, since only then q = ΔH. But the conditions of experiment are those of irreversible process for which q_p is expected to be less than ΔH. Moreover, it is assumed that Cp (Products) are independent of temperature. To get correct result, one should take into account the variation of heat capacities with temperature.

Example 25:

Using bond enthalpies from the following table evaluate ΔH for the isomerization of ethyl alcohol to dimethyl ether :

Bond	*Bond enthalpies at 25°C* $\Delta H.kJ\ mol^{-1}$
H – H	416
O – O	132
O – H	444
C – H	396
C – O	336
C – C	332

Solution:

We have to calculate the enthalpy change of the reaction

$$CH_3CH_2OH \rightarrow CH_3OCH_3$$

The enthalpy change of the reaction will be given as

$\Delta H = \epsilon(C-C) + 5 \times \epsilon(C-H) + \epsilon(C-O) + \epsilon(O-H) - 6 \times \epsilon(C-H) - 2 \times \epsilon(C-O)$

$= \epsilon(C-C) - \epsilon(C-H) - \epsilon(C-O) + \epsilon(O-H) = (332 - 396 - 336 + 444)\ kJ\ mol^{-1} = 44\ kJ\ mol^{-1}$.

Example 26:

Compute the heat of reaction at 1000°C for

$1/2H_2(g) + 1/2\ Cl_2(g) \rightarrow HCL(g) \quad \Delta H^o_{298k} = 92.236\ k\ J\ mol^{-1}$

$C_p^o(H_2,g) = 29.0284 - 0.8355 \times 10^{-3}\ T + 2.0097 \times 10^{-6}\ T^2.$

$C_p^o(Cl_2,g) = 31.6555 + 1.0134 \times 10^{-2}\ T - 4.0337 \times 10^{-6}\ T^2.$

$C_p^o(HCl,g) = 28.1359 + 1.8074 \times 10^{-2}\ T + 1.5453 \times 10^{-6}\ T^2.$

Solution:

$$\Delta H^o_{1273k} = \Delta H^o_{298k} + \int_{298\,k}^{1273\,k} (\Delta C_p)\, dT$$

where $\Delta C_p = \Delta C_p^o$ (HCl, g) – 1/2 C_p^o (H_2, g) – 1/2C_p^o (Cl_2, g)

$$= (28.1359 - 1/2 \times 29.0284 - 1/2 \times 31.6555)$$
$$+ (1.8078 + 1/2 \times 0.8355 - 1/2 \times 10.134) \times 10^{-3}\ T$$
$$+ (1.5453 - 1/2 \times 2.0097 + 1/2 \times 4.0337) \times 10^{-6}\ T^2$$
$$= -2.2061 - 2.8415 \times 10^{-3}\ T + 2.5573 \times 10^{-6}\ T^2$$

Hence,

$$\Delta H^o{}_{1273k} = \left[\begin{array}{l} -92.236\times10^3+(-2.2061)(1273-298)-(2.8415\times10^{-3}) \\ \left(\dfrac{1273^2}{2}-\dfrac{298^3}{2}\right)+(2.5573\times10^{-6})\left(\dfrac{1273^2}{2}-\dfrac{298^3}{2}\right) \\ \end{array}\right]$$

J mol^{-1}

$$= -92236 - 2151.0 - 2176.2 + 1736.0$$
$$= -94827 \text{ J mol}^{-1} = -94.827 \text{ mol}^{-1}$$

Example 27:

The combustion of 1 g of benzene in a bomb calorimeter evolves 41.746 kJ of heat at 25°C, (i) What is ΔU^o for combustion of benzene ? (ii) Calculate heat of formation of benzene.

Solution:

ΔH_f^o (CO_2, g) = – 393.129 kJ mol^{-1}

ΔH_f^o (H_2, l) = – 285.577 kJ mol^{-1}

The molar mass of benzene is 78 g mol^{-1}.

Hence,

$$\Delta U^o = -(41.746 \text{ kJ})\left(\frac{78 \text{ g mol}^{-1}}{1\text{g}}\right)$$
$$= -3256.16 \text{ kJ mol}^{-1}$$

The chemical equation of combustion is

$$C_6H_6(l) + \frac{15}{2} O_2(g) \rightarrow 6CO_2(g) + 3H_2O(l)$$

$$\Delta V_g = 6 - \frac{15}{2} = -\frac{3}{2}$$

Hence,

$$\Delta H^o = \Delta U^o + (\Delta V_g) RT = [(-3256.16 \times 10^3) + (-3/2)(8.314 \times 298)] \text{ J mol}^{-1}$$

$$= -3259.9 \times 10^3 \text{ J mol}^{-1}.$$

Finally, we have

$$\Delta H^o = 6\Delta H_f^o + (CO_2, g) + 3\Delta H_f^o (H_2O, l) - \Delta H_f^o (C_6H_6, l)$$

which gives

$$\Delta H_f^o (C_6H_6, l) = 6\Delta H_f^o + (CO_2, g) + 3\Delta H_f^o (H_2O, l) - \Delta H_f^o$$

$$= [-6 \times 393.129 - 3 \times 285.577 + 3259.9] \text{ kJ mol}^{-1}$$

$$= -44.35 \text{ mol}^{-1}$$

Example 28:

From the following data, calculate ΔH_f at 298.15 k for (i) OH (aq$_l$) and (ii) Na^+(aq) :

ΔH_f^o values is kJ mol^{-1} at 298.15 k for

H_2O,(l) = – 285.8	HCl (aq) = – 167.4
NaOH(aq) = – 469.6	NaCl(aq) = – 407.0

Solution:

(i) $H_2(g) + 1/2\ O_2(g) \rightarrow H_2O(l)$

$$\Delta H^o = -285.8 \text{ kJ mol}^{-1}$$

(ii) $1/2H_2(g) + 1/2\ Cl_2(g) + aq \rightarrow H^+(aq) + Cl^-(aq)$

$$\Delta H^o = -167.4 \text{ kJ mol}^{-1}$$

(iii) $Na(s) + 1/2\ O_2(g) + 1/2\ H_2(g) + aq \rightarrow Na^+(aq) + OH^-(aq)$

$$\Delta H^o = -469.6 \text{ kJ mol}^{-1}$$

(iv) $Na(s) + 1/2\ Cl_2(g) + aq \rightarrow Na^+(aq) + Cl^-(aq)$

$$\Delta H^o = -407.0 \text{ kJ mol}^{-1}$$

Carrying out the manipulation

Eq. (iii) – Eq. (iv) + Eq. (ii)

We get

(v) $1/2O_2(g) + H_2(g) + aq \rightarrow H^+(aq) + OH^-(aq)$

Hence, its enthalpy change becomes

$$\Delta H^o = \Delta H^o_{(m)} - \Delta H^o_{(iv)} + \Delta H^o_{(ii)}$$

$$= (-469.6 + 407.0 - 167.4) \text{ kJ mol}^{-1}$$

$$= -230.0 \text{ kJ mol}^{-1}$$

The enthalpy change of Eq. (v) is also given as

$$\Delta H = \Delta H_f^o(H^+) + \Delta H_f^o(OH^-) - 1/2\ \Delta H_f^o(O_2) - \Delta H_f^o(H_2)$$

$$= 0 + \Delta H_f^o(OH^-) - 0 - 0 = \Delta H_f^o(OH^-)$$

Thus $\quad \Delta H_f^o(OH^-, aq) = -230.0 \text{ kJ mol}^{-1}$

From Eq, (iii), we get

$$\Delta H = \Delta H_f^o(Na^+) + \Delta H_f^o(OH^-) - \Delta H_f^o(Na) - 1/2\ \Delta H_f^o(O_2) - 1/2\ \Delta H_f^o(H_2)$$

$$-469.6 \text{ kJ mol}^{-1} = \Delta H_f^o(Na^+) = -230.0 \text{ kJ mol}^{-1}$$

which gives

$$\Delta H_f^o(Na^+) = (-469.6 + 230.0) \text{ kJ mol}^{-1}$$

$$= -239.6 \text{ kJ mol}^{-1}.$$

2

THE SPECIFIC HEAT OF SOLIDS & GASES

SPECIFIC HEAT OF SOLIDS

Introduction

The specific heat of a substance is defined as the amount of heat required to raise the temperature of 1 gm of the substance through 1°C. The specific heat of a solid increases with rise of temperature and tends to zero as the absolute zero of temperature is approached. For example, diamond (carbon) has a specific heat of 0.095 at 0°C, 0.19 at 100°C and 0.459 at 1000°C. The variation of specific heat of solids with temperature has been explained by a number of scientists. Some of the main attempts are outlined below :

Dulong and Petit's Law

The end of eighteenth century has seen great activity in what we will now call theoretical chemistry. *Lavoisier* and *Laplace* had determined the specific heats of a number of metals with considerable accuracy. In 1819, Dulong and Petit, two French scientists, pointed out that, using the accepted atomic weights for many metals and multiplying them by the specific heats, an approximately constant figure was obtained for each metal. The average value of this constant is 6.4. Thus, Dulong and Petit enunciated their law such as :

"At room temperature, the product of the specific heat and the atomic weight of a solid is a constant/or all solid elements, and is approximately equal to 6.4."

or Atomic weight × Specific Heat = –6.4 (approximately).

The product (atomic weight x specific heat) is termed as atomic heat. Hence, the Dulong and Petit's law may be stated in the alternative form :

"The atomic heats of solid elements, at ordinary temperatures, are constant and approximately equal to 6.4

Successes : When this law was applied to different solid substances, it was found that the atomic heat capacities are almost constant at about 6.2 ± 0.4 cal, per degree in spite of the regular increase in the atomic weights from 7 to 200.

Exceptions : The various exceptions to Dulong and Petit's law are as follows:

(i) According to Dulong and Petit's law, the atomic heats of all solids must be about 6.4. Actually this is not true. For example, the elements beryllium, boron, carbon and silicon have exceptionally low atomic heats: the values at ordinary temperatures being 3.5, 2-5, 1-35 (diamond) and 4.7 cal per degree respectively.

(ii) As the atomic weight of an element is constant, the atomic heat must also be constant, *i.e.*, its value should be same at all temperatures. This is contrary to the observed fact.

For example, diamond shows variation of heat capacity with temperature. This is evident from below :

Temperature	–50°C	107°C	185-5°C	260°C	615°C	808°C
Atomic Heat	076	1.135	2.11	3.28	5.33	5.44

In the case of silicon, the specific heat increases rapidly with temperature and reaches a fairly constant value above 200°C. The variation in metals like lead and tin is very small.

From the above results, one can draw a very important conclusion that the Dulong and Petit's law would apply to all substances provided the temperature is high enough. However, this view is not correct because the heat capacities of those elements which appear to obey the law at ordinary temperatures increases with increasing temperatures. In some cases these values become as high as 9 cal at their melting points. Thus one can safely say that

"There is no one set of conditions under which the Dulong and Petit's law is applicable to alt elements."

(iii) Measurement of specific heats at low temperatures indicates that atomic heat decreases slowly with fall of temperature and below a certain temperature, characteristic of each element, the heat capacity decreases rapidly, tending finally to the zero value at the absolute zero of temperature. This type of change is not permitted by the law of Dulong and Petit. Thus, we can say that :

"The law of Dulong and Petit\ is entirely accidental and the atomic heat capacity of 6 cal. per degree has a theoretical significance only."

Classical Theory

L. Boltzmann (1871) deduced the value of atomic heat of solid from the kinetic theory on the assumption that :

"An ideal solid is composed of atoms vibrating about their respective equilibrium positions but not interacting with one another in any way"

By the above statement it means that in a solid, atoms are allotted fixed positions about which they simply oscillate, possessing kinetic energy of translational as well as potential energy of attraction. The energy of an oscillating atom (harmonic oscillator) is given by

$$E = \text{kinetic energy} + \text{potential energy}$$

$$= \frac{p^2}{2m} + V_{(r)} = \frac{p^2}{2m} + \frac{1}{2} m\omega_o{}^2 x^2 \qquad ...(1)$$

where $V_{(r)}$ is equal to $1/2m\omega_o{}^2x^2$, m is the mass and w_o is the natural frequency of the oscillator. Our next aim is to find out the total energy of the solid due to the various oscillating atoms present in it, *i.e.*, to sum up the energy of all atomic oscillators. The total energy E of a solid is given by

$$E = N \times \overline{E} \qquad ...(2)$$

where E is the mean energy of the atomic oscillator and N, the number of degrees of freedom enjoyed by each atom. From statistical mechanics, it follows that the number of atomic oscillators dN lying between E and E + de at a temperature T is proportional to $e^{-E/KT}$ *i.e.*,

$$dN \propto e^{-E/KT} \quad \text{or } dN = \text{constant} \times e^{-E/KT}$$

$$= \text{constant} \times e^{-(p2/2m\ +\ 1/2mwo2x3)} \qquad ...(3)$$

The mean energy of each oscillator will be given by

$$\overline{E} = \frac{\int_0^\infty E\, dN}{\int_0^\infty dN} \qquad ...(4)$$

Substituting the values of E and dN from equations (1) and (3) in (2), we get

$$\overline{E} = \frac{\int_p \int_x \left(\frac{p^2}{2m} + \frac{1}{2} m\omega_o{}^2 x^2 \right) e^{-p^2/2mKT} \cdot e^{-m\omega_o{}^2 x^2/2mKT} \, dp_x dx}{\int_p \int_x e^{-p^2/2mKT} \cdot e^{-m\omega_o{}^2 x^2/2mKT} \, dp_x dx}$$

or

$$\overline{E} = \frac{\int_{-\infty}^{+\infty} \frac{p^2}{2m} e^{-p^2/2mKT} \, dp_x}{\int_{-\infty}^{+\infty} e^{-p^2/2mKT} \, dp_x} + \frac{\int_{-\infty}^{+\infty} \frac{1}{2} m\omega_o{}^2 x^2 \, e^{-m\omega_o{}^2 x^2/2mKT} \, d\,dx}{\int_{-\infty}^{+\infty} e^{-m\omega_o{}^2 x^2/2mKT} \, dx} \quad ...(5)$$

Using the standard integrals

$$\int_0^\infty e^{-\alpha u^2} \, du = \frac{1}{2} \left(\frac{\pi}{\alpha} \right)^{1/2}$$

and

$$\int_0^\infty u^2 e^{-\alpha u^2} \, du = \frac{1}{4} \left(\frac{\pi}{\alpha^2} \right)^{1/2}$$

we have

$$\frac{\int_{-\infty}^{+\infty} \frac{p^2}{2m} e^{-p^2/2mKT} \, dp_x}{\int_{-\infty}^{+\infty} e^{-p^2/2mKT} \, dp_x} = \frac{2 \times \frac{1}{2m} \times \frac{1}{4} \left[\pi (2mKT)^3 \right]^{1/2}}{2 \times \frac{1}{2} \left[\pi \, (2mKT) \right]^{1/2}} = \frac{KT}{2}$$

$$\frac{\int_{-\infty}^{+\infty} \frac{1}{2} m\omega_o{}^2 x^2 \, /2KT \, dx}{\int_{-\infty}^{+\infty} e^{-m\omega_o{}^2 x^2/2KT} \, dx} = \frac{2 \times \frac{1}{2} \times \frac{1}{4} \left[\pi (2mKT)^3 \right]^{1/2}}{2 \times \frac{1}{2} \left[\pi \, (2KT) \right]^{1/2}} = \frac{KT}{2}$$

Substituting these values in equation (5), we get

$$\overline{E} = \frac{KT}{2} + \frac{KT}{2} = KT \quad ...(6)$$

As an oscillator can vibrate about three mutually perpendicular directions, it means that the mean energy

$$\overline{E} = 3KT \quad ...(7)$$

If there are N atoms in a mole, the to total energy E is obtained by substituting eq. (7) in (2), we get

$$E = 3NKT = 3RT \quad ...(8)$$

Differentiating eq. (7) with respect to temperature, we get

$$\left(\frac{dE}{dT}\right)_V = 3R \text{ or } C_V = 3R \text{ because } C_V = \left(\frac{\partial E}{\partial T}\right)_V$$

or ($C_V = 3 \times 1.98 = 5.94$ cal. per gm. atom per A°)

The above value of 5.94 is nearly equal to the value given by the Dulong and Petit's law. This is the derivation of the law of Dulong and Petit from kinetic theory.

Limitation of Classical Theory

It could not explain the discrepancy of the decrease of specific heat at low temperatures which has been observed in case of all solids. The results are also not in accordance with experiment. The classical derivation and its predication of heat capacity to be independent of temperature are completely disproved.

Einstein Theory of Specific Heat

quantum theory of specific heat. Einstein explained the problem by using the conception of quantum theory. He assumed that the atoms in a solid are all independent, and that each atom acts as simple harmonic oscillator with a common frequency. Thus, the energy of each linear oscillator is given by

$$\in = nh\nu, \; n = 0, 1, 2 \ldots \ldots$$

where n is any positive integer.

The expression for the average energy of an oscillator in quantum theory is quite different from the classical theory. According to Planck's proposal, the linear oscillator vibrates only with integral energy values 0, $\in$, $2\in$, $3\in$, ... where $\in$ represents the elementary quantum, *i.e.*, $\in = h\nu$.

Einstein applied the concept of Planck's theory that atomic system can exist with only certain discrete energies. He further provided the key information about the dependence of heat capacities upon temperature. Before giving the details of Einstein, let us examine the general outline of the calculations. The heat capacity at constant volume is given by

$$C_V = \Delta E / \Delta T$$

where ΔE is the change in the total energy of a mole of substance produced by the temperature change, ΔT. We shall calculate ΔE from the expression

$$\Delta E = \Delta N \overline{\in}$$

where N is the number of atoms in a crystal, and $\Delta \overline{\in}$ is the change in the average energy of an atom in the crystal produced by the temperature increase ΔT. Our first objective then is to find an expression that gives us information how the energy $\overline{\in}$ depends on temperature.

Let us consider (as Einstein did) a mole of monoatomic solid which consists of N mass points that can oscillate in the three mutually perpendicular directions with a vibrational frequency ν. According to Planck's quantized energy, an atom oscillating in one direction could have only one set of the energies given by

$$\in_n = nh\nu, \; n = 0, 1, 2, 3, \; \ldots \ldots \qquad \ldots(1)$$

where h is Planck's constant and is a proportionality factor between frequency and energy, and n is an integer. Planck's constant has the value 6.6×10^{-27} ergs and the frequency ν is equal to 10^{12} sec^{-1} for many solids. Although an actual atom in a solid is a three-dimensional oscillator, yet we assume it to be a one-dimensional oscillator and we will correct it for this discrepancy later. At thermal equilibrium, the oscillators in a crystal will be distributed among the various allowed energies according to the Boltzmann distribution law which states that the number N_n of atoms with energy $\in_n$ is related ti the number N_o with energy $\in_o = 0$ by the expression

$$N_n = N_o \, e^{-\in n/KT} = N_o \, e^{-nh\nu/KT} \qquad \ldots(2)$$

The total number of particles, N, is equal to the sum of the number in each energy state,

$$N = N_o + N_1 + N_2 + eN_3 + N_4 + \ldots$$

Substitution of the Boltzmann factor, eq. (2) gives

$$N = N_o + N_o e^{-h\nu/KT} + No \; e^{-2h\nu/KT} + \ldots$$

$$= N_o \sum_{n=o}^{\infty} e^{-nh\nu/KT}$$

We will not calculate the total energy due to the oscillation of the atoms in this one direction only. We can do this by multiplying energy $\in_o$ by the number of particles which have that energy, and adding all these quantities :

$$\Delta E = \in_o N_O + \in_1 N_1 + \in_2 N_2 + \ldots$$

$$= 0N_O + h\nu N_1 + 2h\nu N_2 + \ldots$$

$$= 0 + h\nu N_o e^{-h\nu/KT} + 2h\nu N_o e{-}2h\nu/kT + \ldots$$

$$= N_o \sum_1^{\infty} nh\nu\, e - nh\nu / kT$$

But $$\overline{\in} = \frac{\Delta E}{N} = \frac{h\nu \sum ne^{-nh\nu/kT}}{\sum e^{-nh\nu/kT}} \quad \ldots(3)$$

where $\overline{\in}$ is the average energy of a one-dimensional oscillator. In order to evaluate the value of $\overline{\in}$, the summation in eq. (3) must be done. This can be done if we make the substitution

$$y = e^{-h\nu/KT}\text{, which gives } e^{-nh\nu/KT} = y^n$$

Then, for the series in the denominator of Eq. (3), we get

$$\sum_{n=1}^{\infty} e^{-nh\nu/kT} = \sum_{n=1}^{\infty} y^n = 1 + y^2 + y^3 + \ldots = \frac{1}{1-y} \quad \ldots(3A)$$

This last step can be verified by algebraic long division of 12 by 1 – y. We treat the series in the number at or of Eq. (3) in the same manner

$$\sum_1^{\infty} ne^{-nh\nu/kT} = \sum_1^{\infty} ny^n = y\left(1 + 2y + 3y^2 + \ldots\right) = \frac{y}{(1-y)^2} \quad \ldots(3B)$$

Substituting equation (3A) and (3B) in (3), we get

$$\overline{\in} = \frac{h\nu y}{1-y} = \frac{h\nu e^{-h\nu/kT}}{1-e^{-h\nu/kT}}$$

Multiplying numerator and denominator by $e^{h\nu/kT}$ yields

$$\overline{\in} = \frac{h\nu}{e^{h\nu/kT}-1}$$

This is the average energy of a one-dimension oscillator. The average energy of an atom in a crystal is three times this amount since the atom vibrates in three directions.

Thus, $$\overline{\in}_{atom} = \frac{3h\nu}{e^{h\nu/kT}-1} \quad \ldots(4)$$

We will not compute the heat capacity of the solid at such a high temperature so that $h\nu/kT << 1$. Under these conditions, the denominator of Eq. (4) simplifies considerably if we apply the fundamental properties of the exponential function.

$$e^{h\nu/kT} = 1 + \frac{h\nu}{kT} + \frac{1}{2}\left(\frac{h\nu}{kT}\right)^2 + \frac{1}{6}\left(\frac{h\nu}{kT}\right)^3 ... +$$

But $h\nu/kT << 1$, all terms after the second are very small and may be neglected in comparison with $1 + h\nu/kT$. Then, we get

$$e^{h\nu/kT} \approx 1 + \frac{h\nu}{kT} \qquad ...(5)$$

which when substituted into eq. (4) gives

$$\overline{\in} = \frac{3h\nu}{e^{h\nu/kT} - 1} = \frac{3h\nu}{h\nu / kT} = 3kT \qquad ...96)$$

Therefore, the total energy of one mole of atoms is given by

$$\Delta E = \overline{N} \in = 3NkT \qquad [\backslash\ R = Nk] \qquad ...(7)$$

and the heat capacity is

$$C_V = \Delta E / \Delta T = 3R = 6 \text{ cal/ mole/ deg} \qquad ...(8)$$

The result (eq. 8) is very close to the value used in Dulong and Petit's law. The Einstein's theory thus explains specific heat of solids at high temperatures only. This has been found to be true for a large number of elements. It does not hold true in the case of elements where ν is sufficiently large. In these cases, a very high temperature is needed to attain the classical value. It means that is much depends upon the variable ν. The plot of C_V against T clearly shows that for low values of ν, it holds true (Fig. 2.1.)

It is more difficult to explain the behaviour of heat capacity at low temperatures by Einstein theory, because it is not possible to simplify equation (3) appreciably when T is very small. However, if we do simplification of equation (3) at low temperatures by making serious approximations, we observe tat the specific heat of solid decreases as the temperature falls and tends to be zero at absolute zero.

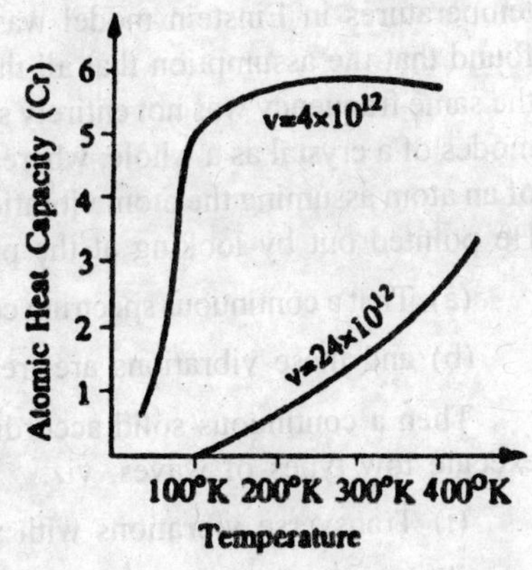

Fig. 2.1.

Successes of the Einstein Theory :

1. Einstein explained theoretically that the specific heat (atomic heat) increase with temperature and tends to the values given by Dulong and Petit at high temperatures. For some substances even at ordinary temperatures the condition is satisfied.
2. Specific heat decreases as the temperature falls and tends to be zero at absolute zero.

Limitations of Einstein Theory

1. Einstein theory is only approximate and its limitations are more serious at low temperatures. It was observed that in case of elements like copper, aluminium etc., the atomic heats at low temperatures decrease more rapidly than that predicted by Einstein theory.
2. Another weakness of Einstein theory is that while deducing his theory ν was used. This is obtained empirically and cannot be verified from any other independent physical data.
3. Einstein assumed that the vibrations of a particular atom must be of the same frequency. But the vibrations of a particular atom must be very complex, because it is under the field of force of other vibrating atoms.

DEBYE THEORY

The discrepancy in theoretical and experimental results at low the temperatures in Einstein model was recognised by Debye (1912). He found that the assumption that all the particles of a crystal vibrate with the same frequency was not entirely sound. He considered the vibrational modes of a crystal as a whole, whereas Einstein considered the vibration of an atom assuming the atom vibrations to be independent of each other. He pointed out by looking at the problem :

(a) That a continuous spectrum could result from modes of vibrations.

(b) and these vibrations are treated as elastic-type vibrations.

Then a continuous solid according to the above conception would execute tow types of waves, viz.

(i) Transverse vibrations with velocity $C_t = \sqrt{(n/\rho)}$

(ii) Longitudinal vibrations with velocity $C_t = \sqrt{(k + 4/3\ n/\rho)}$

Here n is rigidity coefficient and k is the bulk of modulus. The number of possible vibrational modes in second case enclosed in volume V in frequency internal ν and ν + dν is given by

$$\frac{4\pi\nu^2}{C_l^{\,3}} d_\nu V \qquad ...(9)$$

and the number of modes of vibration in first case would be (in same frequency interval) given by

$$= \frac{8\pi\nu^2 d\nu}{C_l^{\,3}} V \qquad ...10)$$

because in this case there would be two waves polarised at right angles to each other.

Hence for an elastic sold which can transmit three independent vibrations in any given direction, namely, two transverse vibrations polarised in planes at right angles to each other and a longitudinal wave is given by

$$\int_0^{\nu_m} 4\pi V\left(\frac{1}{C_t^{\,3}} + \frac{2}{C_t^{\,3}}\right)\nu^2\, d\nu \qquad ...(11)$$

Now, it is very clear that in one gram, *i.e.*, Number of atoms are not vibrating with the same frequency but this is system of 3N oscillators. Hence,

$$3N = \int_0^{\nu_m} 4\pi V\left(\frac{1}{C_t^{\,3}} + \frac{2}{C_t^{\,3}}\right)\nu^2\, d\nu \qquad ...(11A)$$

or

$$\nu_m^{\,3} = \frac{9N}{4\pi V\left(\frac{1}{C_t^{\,3}} + \frac{2}{C_t^{\,3}}\right)} \qquad ...(12)$$

The energy of a harmonic oscillators, according to Planck, is given by

$$\frac{h\nu}{\left(e^{h\nu/kT} - 1\right)}$$

So the energy for 3N harmonic oscillators is given as

$$E = 3N\frac{h\nu}{\left(e^{h\nu/kT} - 1\right)}$$

Now this equation assumes the term from equation (11A)

$$E = 4\pi V\left(\frac{1}{C_t^3} + \frac{2}{C_t^3}\right)\int_0^{\nu_m} \frac{h\nu^3}{\left(e^{h\nu/kT} - 1\right)} d\nu \quad ...(13)$$

From equation 12), we have

$$E = \frac{9N}{\nu_m^3}\int_0^{\nu_m} \frac{h\nu^3\, d\nu}{e^{h\nu/kT} - 1} \quad ...(14)$$

Differentiating the above with respect to temperature, we obtain

$$C_V = \frac{9R}{\nu_m^3}\int_0^{\nu_m} \frac{e^{h\nu/kT}}{\left(e^{h\nu/kT} - 1\right)^2}\left(\frac{h\nu}{kT}\right)^3 \nu^2\, d\nu \quad ...(15)$$

Equations (15) is known as Debye heat capacity equation. To evaluate the integral, put

$$\frac{h\nu}{kT} = \Psi\,;\ \frac{h\nu_m}{k} = \theta\,;\ \frac{\theta}{T} = \Psi_m = x$$

Substituting these values in equation (15), we get

$$C_V = 3R\left[\frac{12}{x^3}\int_0^x \frac{\Psi^3\, d\Psi}{e\Psi - 1} - \frac{3x}{e^x - 1}\right] \quad ...(16)$$

Equation (16) is known as Debye's expression for the atomic heat of a monoatomic solid. The term within the square bracket is a function of x and is generally designated by Debye function, D(x). It s value has been calculated for the values of x which range from to ∞.

Limiting Case :

At moderate and very high temperatures, equation (16) becomes as

$$C_V = 3R\left[1 - \frac{1}{20}\left(\frac{\theta}{T}\right)^2 + \frac{1}{560}\left(\frac{\theta}{T}\right)^4 - ...\right] \quad ...(17)$$

At sufficiently high temperatures, θ/T becomes small enough for all the terms in the brackets of equation (17) other than the first to be neglected; C_V then equal to the classical value 3R.

(ii) At very low temperatures T is very small and x and Ψ are very large. The second term in equation (16), becomes negligible. Also, the denominator of the first term is also large unless ν is small. At low temperatures, θ/T tends to ∞. Hence, one can replace the upper limit of the first expression of equation (16) be ∞. We have thus

$$C_v = 3R\,\frac{12}{x^3}\int_0^{\infty}\frac{\Psi^3\,d\Psi}{e^{\Psi}-1}$$

$$= 3R\,\frac{12\pi^4}{x^3\,15} = 3R.\frac{12\pi^4}{15}.\left(\frac{T}{\theta}\right)^3 = 77.94\times 3R\left(\frac{T}{\theta}\right)^3$$

$$= 464.4\left(\frac{T}{\theta}\right)^3 \text{ cal/deg/gm atom.} \qquad ...(18)$$

As θ is constant for a substance, the specific heat at extremely low temperatures varies as T^3. This statement is known as Debye's – T^3 law. This law has been verified in a number of cases. When t = 0° A, equation (18) becomes as $C_V = 0$

i.e., atomic heat capacity vanishes at absolute zero.

Calculation of Characteristic Temperatures :

we know that

$$\theta = \frac{h\nu_m}{k} = \frac{h}{k}\left\{\frac{9N}{4\pi\nu\left(\frac{2}{C_t^{\,3}}+\frac{1}{C_t^{\,3}}\right)}\right\}^{\frac{1}{2}} \qquad \text{[From eq. (12)]} \qquad ...(19)$$

and $$C_t = \left[\sqrt{\left(k+\frac{4}{3}n\right)}\right]\Big/\rho \text{ and } C_t = \sqrt{(n/\rho)} \qquad ...(20)$$

$$\therefore \qquad \theta = \frac{h}{k}\left(\frac{9N}{4\pi V}\right)^{1/3}\rho^{1/2}\left[\frac{2}{n^{3/2}}+\frac{2}{(k+4/3)^{3/2}}\right]^{1/2} \qquad ...(21)$$

where all the letters have their usual significance.

Hence the value of θ can be calculated from equation (21).

The Debye theory predicts that the heat capacity of a solid as a function of temperature should depend only on the characteristic frequency, ν_m. These crystals with closely spaced levels, as a result of weak bonds or heavy atoms, as in lead reach the classical heat capacity value at low temperatures while the crystals with large vibrational spacings reach the value only at high temperatures.

Weaknesses of Debye Theory

(i) This theory does not agree with its theoretical foundation.

(ii) Debye theory cannot be extend to low temperature.

Criticism of Debye Theory

(i) This theory explains quite satisfactorily the aviation of specific heat of solids, but it does not agree with its theoretical functions. It was assumed that :

(ii) When a solid is heated, it is supposed that elastic vibrations like those of continuous medium exist within the body having frequencies from 0 to ν_m, where the spectrum guillotined sharply.

Even if a solid be isotropic, it cannot be treated as continuous since it is composed of atoms separated in space.

(iii) Here the solid is considered to be an isotropic body, but in reality solids are made of definite crystals, either in the form of single crystals or minute crystals brought together by pressure. It has not considered crystalline structure in its derivation.

Slater inferred that equation (12) becomes incorrect due to two reasons :

(a) That the velocity of elastic vibrations depends upon the wavelength of the vibration when it is comparable with atomic distances (as it is for frequencies approaching n_m) and hence the change of velocity with frequency must be taken into account in deriving Eq. (12).

(b) In an actual solid composed of N atoms, one has to find out the normal or independent modes of vibration of atoms in the solid. This problem was solved by Born by assuming "periodic" boundary conditions or assuming fixed boundaries, the force between atoms to be given by Hooke's law. He showed that there will 3N normal modes of vibration. Thus the total number of modes comes out to be the same as the number of degree of freedom of solid as assumed by Born which is different from Debye even for anisotropic medium.

The Debye model, as also the Einstein's, is applicable to assemblies of harmonic oscillators and cannot, therefore, deal with properties of solids which arise primarily due to anharmonicity of lattice vibrations. Thus to improve upon the Debye theory of specific heat due to lattice vibrations, a more precise vibrational spectrum is required. The problem in its three-dimensional form is complicated. Although to explain the basis and nature of the anharmonic effect, a one-dimensional picture may be constructed in a rather simple manner to understand the problem.

Consider a one-dimensional lattice having a 'grating space' equal to s and containing n identical atoms in it. If the displacement of, say, ith atom from the mean position is x_i the energy E of the lattice is

$$E = \sum_{i-1}^{n} \frac{1}{2} m x_i^{\,2} + \sum_{i-1}^{n} \frac{1}{2} f\left[x_i - x_{i-1}\right]^2 \qquad ...(22)$$

where n represents the number of atoms each of mass m in the lattice and f the force constant. The Lagrangian equation of motion, considering the ith atom, are as follows :

$$m z_i = f[x_{i+1}\ x_i - (x_i - x_{i-1})] = f[x_{i+1} + x_{i-1} - 2x_i], \qquad ...(23)$$

$$i = 1, 2, ..., n$$

If the linear crystal lattice is imagined to be bent into a circle (endless chain) such that the nth atom is next tot the first atom, then the so-called 'cyclic boundary conditions' will be obviously

$$x_i(t) = x_{i+n}(t) \qquad ...(24)$$

where t denotes time motion,

The solutions of Lagrange's equations may be expressed as

$$x_i(t) = A e^{j(\omega t + i\phi)} \qquad ...(25)$$

where $j = (-1)^{1/2}$, A is a constant, $\omega = 2\pi\nu$ and

$$\phi k = \frac{2\pi k}{n}, \; k = 1, 2, ... n \qquad ...(26)$$

Hence from eq. (25)

$$x_i(t) = A e^{j\omega t} e^{2\pi i j/n} \qquad ...(27)$$

Now substitute eq (25) in eq. (23)

$$-m\omega^2 = f\left[e^{j\phi} + e^{-j\phi} - 2\right] = f\left[e^{j\phi/2} + e^{-j\phi/2}\right]^2$$

or $\qquad m\omega^2\ 4\ f \sin^2 \phi/2 \qquad ...(28)$

And since $\phi = 2\pi k/n$,

$$\omega k = 2\left(\frac{f}{m}\right)^{\frac{1}{2}} \sin\frac{\pi k}{n'} k = 1, 2 \ldots, n \qquad \ldots(29)$$

and
$$V_k = \frac{\omega k}{2\pi} = \frac{1}{\pi}\left(\frac{f}{m}\right)^{1/2} \sin\frac{\pi k}{n} \qquad \ldots(30)$$

Differentiating eq. (30), we get

$$dv_k = \frac{1}{\pi}\left(\frac{f}{m}\right)^{1/2} \sin\frac{\pi k}{n} dk \qquad \ldots(31)$$

Now as $\sin \phi = \sin (\pi - \phi)$, for $\phi \leq \pi/2$; thus there will be two physically distinct values ϕ and $(\pi - \phi)$ which yield the same value of $\sin \phi$, Hence for each value of k such that $1 \leq k \leq n/2$, there will be two distinct normal vibrations, corresponding to $\phi_k = 2\pi k/n$ and $\phi_{n-k} = \frac{2\pi(n-k)}{n}$. But these modes have frequency νk as given by eq. (30), The number of frequencies between ν and $\nu + d\nu$ may also be defined as $\frac{1}{d\nu_k / dk} = g(\nu)$ – the number of vibrational modes, obtained as :

$$g(\nu)\, d\nu = \frac{2n}{\cos(\pi k / n)} (m/f)^{1/2} d\nu \qquad \ldots(32)$$

The additional factor 2 in eq. (32) is present because for every value of k there are two distinct frequencies.

Also denoting

$$\nu_o = \frac{1}{\pi}\left(\frac{f}{m}\right)^{1/2} \qquad \ldots(33)$$

We have $\quad \nu = \nu_o \sin\frac{\pi k}{n}$

Such that
$$\cos\frac{nk}{n} = \frac{\left(\nu_o^2 - \nu^2\right)^{1/2}}{\nu_o} \qquad \ldots(34)$$

We have, substituting eq. (34) in eq. (32),

$$g(\nu)\, d\nu = \frac{2n}{\pi} \frac{d\nu}{\left(\nu_o^2 - \nu^2\right)^{1/2}} \qquad \ldots(35)$$

Eq. (35) thus gives the frequency spectrum for a non-dimensional linear lattice. This should be compared with the Debye spectrum which for a one-dimensional oscillator, is as :

$$g(\nu)\,d\nu = \nu\,d\nu. \qquad ...(36)$$

Eq. (36) therefore gives an idea of the true picture of the normal vibrations of the lattice and one observes that the spectrum differs considerably from the Debye spectrums.

An interesting calculation, according to Born theory, was given by Thirring. He pointed out that the specific heat at high temperatures could be worked out without finding an explicit form for the spectrum and without any further approximations. This method is to expand the mean energy of an oscillator in powers of hν/kT and the result is

$$E - kT\left\{1 - \sum_{n-1}^{\infty}(-1)^n \frac{B_n}{(2n)!}\left(\frac{h\nu}{kT}\right)^{2n}\right\}$$

where B_n are the Bernoulli numbers.

The heat capacity of the linear oscillator is then given by

$$C_V = \frac{dE}{dT} = k\left\{1 + \sum_{n-1}^{\infty}(-1)^n \frac{B_n(2n-1)}{(2n)!}\left(\frac{h\nu}{kT}\right)2n\right\}$$

Born and V. Karman theory depends on the calculation of frequency spectrum as a dynamic crystal lattice.

It is expected to apply these theories only to a strictly isotropic body, and to one in which all the atoms are the same. It has been extended by Born and others to apply this theory to crystal and to bodies containing more than one type of atoms. These theories take into account the fact that body is no continuous but consists of atoms, arranged in a lattice.

BLACKMAN'S LATTICE THEORY OF SPECIFIC HEATS

It is clear that the theory which has been developed so far gives a general account of specific heat of solids except at low temperatures.

From the results obtained by Debye theory, it appears that it is unsatisfactory to treat crystal as continuum. The vibrational spectrum of a crystal lattice can differ very much from that of an elastic solid. Calculation of the vibrational spectrum of a two dimensional lattice by Blackman showed that it differed greatly from the assumed by Debye.

Blackman took into account of the interacations between the closed unneighborly particles of the lattice and also between next nearest neighbour particles.

According to this theory specific heats at high and at low temperatures depend on the different combinations of the forces between particles. In high temperature region, the specific heap is controlled by the normal vibrations of the crystals, and the low frequency end of the spectrum plays a negligible part in determining it. On the other hand the specific heat at low temperatures is controlled by the low frequency end of the spectrum.

In order to follow the details of the variation of the specific heat of solid with temperature, spectrum in each case ins recorded.

Comparison with Debye's Theory

(i) Debye theory is remarkable in view of complications occurring in the lattice theory though Debye theory fails miserably at low temperatures.

(ii) the calculation of the vibrational spectrum has thrown new light on the theory of specific heat of solids. The vibrational spectrums is of quite different character from that assumed by Debye.

SPECIFIC HEAT OF GASES

The specific heat of a substance is defined as the amount of heat required to raise the temperature of unit mass of that substances through 1°C .

Generally, the heat supplied to the substance is used up in increasing the temperature and also in performing external work.

Thus, the specific heat of a substance may be spoken with definiteness only when the conditions under which the heating takes place are defined.

In case of solids and liquids, the change of volume and hence the external work during the change of temperature is very small and hence there will be only one specific heat in case of solids and liquids.

In case of gases, there occurs big changes of both pressure and volume due to change in temperature.

Therefore, it becomes important to define the conditions under which the heat is applied, *i.e.*, either the pressure of the volume may be kept constant while temperature rises. Therefore, a gas has two specific heats :

(a) Specific Heat of a Gas at Constant Volume

It is the amount of heat required to raise temperature of one gm of the gas through 1°C, the volume being kept constant. It is denoted by C_V.

(b) Specific Heat of a Gas at Constant Pressure

It is the amount of heat required to raise the temperature of one gm of the gas through 1°C, the pressure being kept constant. It is denoted by C_P.

When a gas is heated at constant volume, the whole heat supplied is used up to raise its temperature. On the other hand, if the gas is heated at constant pressure, a part of heat is used increase its temperature and a part is used to do external work to keep the pressure constant *i.e.*,

$$C_P = C_V + \text{work done.}$$

3

NON-EQUILIBRIUM AND CLASSICAL THERMODYNAMICS

NON-EQUILIBRIUM THERMODYNAMICS

Introduction

Many transport processes may take place in the system which may be irreversible and the system may not be in equilibrium with respect to these quantities. A few examples of such transport phenomena include flow of heal along a metal bar whose ends are kept at a fixed but at different temperatures, the development of heat when an el^ctric current flows through a metallic conductor, diffusion of a solid or a fluid across a concentration gradient, etc. The branch of science dealing with the study of thermodynamic properties of the systems which are not in equilibrium and Involve transport processes which are irreversible is termed as Non-equilibrium thermodynamics for Irreversible processes. In order to develop the general theory for the irreversible processes, some additional postulates (assumptions) have to be introduced. These postulates are solely applicable only when the system is "not too far" from equilibrium *i.e.* to the states which are close to equilibrium. Hence, irreversible thermodynamics is applicable to those systems which are not too far from equilibrium.

THEOREM OF WIENER-KHINTCHINE

This theorem is concerned with a random process.

A random process may be defined as a process in which a variable y is not dependent in a completely definite way on the independent time variable t. Two important characteristics of a random process are as follows :

(i) The power spectrum of the process, and

(ii) The correlation function of the process.

The Wiener-Khintchine theorem gives a relationship between these two characteristics, (i) and (ii).

We will now consider a periodic variable y (t) which could be expanded in Fourier series in the interval $0 < t < T$ as given below :

$$y(t) = \alpha_o + \sum_{n=1}^{\infty} \alpha_n \cos(2\pi f_n t) + \sum_{n=1}^{\infty} b_n \sin(2\pi f_n t). \quad ...(1)$$

In the above equation $f_n = n/T$, time period $T = 1/f_o$; f_u = fundamental frequency and $f_n = nf_o$.

Hence we assume that <y (t)> = 0. Where the angular parentheses, > denotes time average. As the average is regarded to be zero, there will be no constant term in Fourier series. Under this condition, we get

$$y(t) = \sum_{n=1}^{\infty} \{\alpha_n \cos(2\pi f_n t) + b_n \sin(2\pi f_n t)\} \quad ...(2)$$

Here $a_n = \frac{2}{T}\int_o^T y(t)\cos(2\pi f_n t)dt$ ...93)

and $b_n = \frac{2}{T}\int_o^T y(t)\sin(2\pi f_n t)dt$

Before we derive the theorem, we will prove that the ensemble average of the time average power dissipation $\overline{\langle P_n \rangle}$ associated with n[th] components of y (t) has been equal to the square of standard deviation σ_n, *i.e.*,

$$\overline{\langle P_n \rangle} = \sigma_n^2. \quad ...(4)$$

The power in the nth component may be given as follows :

$$P_n = \{a_n \cos(2\pi f_n t) + b_n \sin(2\pi f_n t)\}^2 \quad ...(5)$$

The time average of P_n may be given as follows :

$$<P_n> = \frac{1}{2} < a_n^2 > + \frac{1}{2} < b_n^2 > = < a_n^2 + b_n^2 > \frac{1}{2}$$

because $<\cos^2(2\pi f_n t)> = \frac{1}{2}$, $< \sin^2(2\pi f_n t)> = \frac{1}{2}$

and $<\cos(2\pi f_n t). \sin(2\pi f_n t)> = 0$

Now we regard the ensemble average (which is an average over a large set of independent records for random process) which is denoted here by a over the quantity. We get

$$\overline{a_n} = 0, \overline{b_n} = 0, \overline{a_n b_m} = 0 \qquad \text{...(6)}$$

$$\overline{a_n a_m} = \overline{b_n b_m} = \sigma_n^2 \delta_{nm}$$

where σ_n refers to standard deviation.

Now $\overline{\langle P_n \rangle} = \overline{\{a_n \cos(2\pi f_n t) + b_n \sin(2\pi f_n t)\}^2}$

$$= \sigma_n^2 \{\cos^2(2\pi f_n t) + \sin^2(2\pi f_n t)\}$$

$$= \sigma_n^2$$

$$\therefore \quad \overline{\langle P_n \rangle} = \sigma_n^2. \qquad \text{...(7)}$$

Power Spectrum

One can define the power spectrum or spectral density W (f) of random process as the ensemble average of the time average of power dissipation per unit frequency bandwidth. If we select a frequency bandwidth Δf_n equal to the separation between two adjacent frequencies that is

$$\Delta f_n = f_{n+1} - f_n = \frac{n+1}{T} - \frac{n}{T} = \frac{1}{T},$$

We get $W(f_n)\Delta f_n = \overline{\langle \rho_n \rangle} = \sigma_n^2 \qquad \text{...(8)}$

It can be proved that

$$\overline{y^2(t)} = \sum_n \sigma_n^2$$

or

$$\overline{y^2(t)} = \sum_n W(f_n) = \int_0^\infty W(f)\,df \qquad \text{...(9)}$$

The integral of the power spectrum over all frequencies provides the ensemble average total power, which is assumed to be independent of time, so we talk of it simply as the average total power.

Correlation Function

The correlation function may be defined as follows :

$$K(\pi) = \langle y(t)\, y(t + \pi) \rangle$$

Without altering the result let us take an ensemble average of the time average <y (t) y (t + π)>.

So that,

$$K(\pi) = \overline{\langle y(t)\, y\,(t+\pi)\rangle}$$

$$\therefore \quad K(\pi) = \overline{\left\langle \sum_{n,\,m}\left[a_n \cos(2\pi\, f_n t)\, b_n \cos(2\pi\, f_n t)\right]\left[a_n \cos 2\pi\, f_m (t+\pi)\right.\right.}$$

$$\overline{\left.\left. + \, b_m \sin 2\pi\, f_m\,(t+\pi)\right]\right\rangle}$$

$$= \frac{1}{2}\sum_n \overline{\left(a_n{}^2 + b_n{}^2\right)} \cos\,(2\pi\, f_n\, \pi) = \sum_n \sigma_n{}^2 \cos\,(2\pi\, f\, \pi)\,df$$

$$= \int_0^\infty W\,(f)\cos\,(2\pi f\,\pi)\;df$$

Hence $\quad K\,(\pi) = \int_0^\infty W\,(f)\cos\,(2\pi f\,\pi)\;df \qquad \text{...(10)}$

or $\quad W\,(f) = 4\int_0^\infty K\,(\pi)\cos\,(2\pi f\,\pi)\;d\pi$

(Using Fourier transform) ...(11)

Equations (10) and (11) are termed as Wiener-Khintchine theorem.

PRODUCTION OF THE ENTROPY

The entropy of an isolated system in equilibrium is maximum. Hence, if such a system is not in equilibrium, the entropy will increase but may not decrease *i.e.*, equilibrium lies in the direction of increasing entropy. This is termed "entropy production.

The concept of entropy production in an irreversible process may be understood in a simple manner as follows. It is known that

$$dS \geq \frac{dQ}{T}$$

or which can be rearranged in the form

$$dS - \frac{dQ}{T} \geq 0$$

The quantity on the left is greater than or equal to zero. So we may write

$$dS - \frac{dQ}{T} = d\sigma \qquad \text{...(1)}$$

where, $d\sigma$ will be either zero or positive.

If it is assumed that the system is in contact with a reservoir at T, and a quantity of heat dQ flows into the system, then a quantity, –dQ, flows in the reservoir. If the quantity, dQ, is transferred reversibly to the reservoir, then the entropy change of the reservoir is

$$dS = -\frac{dQ}{T} \qquad \text{...(2)}$$

so that the Eq. (1) can be written as

$$dS + dS_{res} = d\sigma \qquad \text{...(3)}$$

The quantity $d\sigma$ refers to the entropy increase of the system plus that of the surroundings the reservoir) The $d\sigma$ is called the *entropy production* of the process. For an irreversible process entropy production is positive while for a reversible process, the entropy production is zero.

Let us now derive the exact expressions for the entropy production accompanying the flow of heat or flow of matter (which are irreversible processes) for the systems not in equilibrium but very close to the equilibrium.

(a) *Entropy Production in Matter Flow :* Consider thc flow of δn moles of some components of the matter (or of electricity) from one region of temperature T_1. where the chemical potential is μ_1, to a second region where the temperature is T_2 and the chemical potential is μ_2.

When matter (fluid) flows from a cooler region to the hotter region, not only the het content of the substance increases say be dH per mole (or represented as $\overline{H}$) but also a heat effect takes place when the fluid enters the surfαce of the barrier and passes into the interior. This may be thought of as primarily the heat of solution. This quantity is called the heat of transfer and is written as Q* per mole.

Now the partial molal entropy of a system (entropy per mole) having molal enthalpy $\overline{H}$ and chemical potential μ and maintained at temperature T is given by

$$\overline{S} = \frac{\overline{H}}{T} - \frac{\mu}{T} \qquad \text{...(8)}$$

Taking that het of transfer (Q*) also into consideration the above equation becomes

$$\overline{S} = \frac{(\overline{H} + Q^*)}{T} - \frac{\mu}{T} \qquad \text{...(9)}$$

According to first law of thermodynamics, the sum $\overline{H} + Q^*$ must be same whether we are considering region 1 or region 2.

Thus the partial molal entropy of region 1 is

$$\overline{S}_1 = \frac{(\overline{H} + Q^*)}{T_1} - \frac{\mu_1}{T_2} \qquad \text{...(10)}$$

and the partial molal entropy of region 2 is

$$\overline{S}_2 = \frac{(\overline{H} + Q^*)}{T_2} - \frac{\mu_2}{T_2} \qquad \text{...(11)}$$

∴ Increase of entropy per mole transfer of matter from region 1 to region 2 will be

$$\overline{dS} = \overline{S}_2 - \overline{S}_1 = \left(\frac{\mu_1}{T_1} - \frac{\mu_2}{T_2}\right) + (\overline{H} + Q^*)\left(\frac{1}{T_2} - \frac{1}{T_1}\right) \qquad \text{...(12)}$$

∴ Entropy increase for the transfer of δn moles will be given by

$$dS_{irr} = \delta n\left[\left(\frac{\mu_1}{T_1} - \frac{\mu_2}{T_2}\right) + (\overline{H} + Q^*)\left(\frac{1}{T_2} - \frac{1}{T_1}\right)\right] \qquad \text{...(13)}$$

or for infinitesimal differences, as before, we can write

$$dS_{irr} = \delta n\left[-d\left(\frac{\mu}{T}\right) + (\overline{H} + Q^*)\, d\left(\frac{1}{T}\right)\right]$$

If $T_2 = T_1$, the second term in eqn. (13) and hence in eqn. (14) drops out . This explains why the heat of transfer never needs to be considered in isothermal problems.

(b) *Entropy Production in Heat Flow* : Suppose our system is having a metal rod. Consider two regions of the metal rod maintained at different temperatures T_1 and T_2 *i.e.*, each region is isothermal. This can be achieved by allowing the heat to enter one region and leave the other region at the same rate.

If δq_1 refers to the amount of heat that enters the region at T_1 per unit time and δq_2 is the amount of heat that leaves the region at T_2 per unit time, then

$$\text{Entropy change in region } 1 = \frac{\delta q_1}{T_1}$$

and entropy change in region $2 = \frac{\delta q_2}{T_2}$

But as each of the regions 1 and 2 is isothermal, therefore we must have

$$\delta q_1 = -\delta q_2 \qquad ...(4)$$

∴ Net entropy change accompanying the irreverible flow of heat will be given be

$$dS_{irr} = \delta q_2 \left(\frac{1}{T_2} - \frac{1}{T_1}\right) = \delta q_2 \frac{T_1 - T_2}{T_1 T_2} \qquad ...(5)$$

The irreversible entropy increase (dS_{irr}) must, of course, be positive, Hence $T_2 < T_1$ which has been found to be in agreement with the well known direction of spontaneous heat flow.

If the temperature difference is infinitesimal (so that $T_1 - T_2 = dT$ and $T_1 = T_2 = T$ say so that $T_1T_2 = T_2$), the above equation becomes

$$dS_{irr} = \delta q \frac{dT}{T^2} \qquad ...(6)$$

which can be written in the form

$$dS_{irr} = d\ q\ d(1/T) \qquad ...(7)$$

ASSUMPTION OF LOCAL EQUILIBRIUM

This assumption suppose the system is divided into a number of small volume elements (called cells or sub-systems). Every cell will have a fixed volume V. It is further assumed that at any instant of time, the subsystem has an internal energy U and contains m_i mass units of the molecular species i. At equilibrium, temperature T, pressure P. partial specific Gibb's function ($\mu_i = (\partial G/\partial m_i)_{T,P}$ and entropy S of a subsystem are well defined parameters and because V is constant, these parameters are solely determined by U and m_i for the subsystem (note that here μ_i stands for the chemical potential per unit mass whereas ordinarily this symbol represents the chemical potential per mole).

If equilibrium does not exist, it becomes necessary to redefine the temperature, pressure, partial specific Gibbs function and entropy. It is assumed that

"Temperature (T), pressure (P), partial specific Gibbs function (G) and entropy S for a cell in a non-equilibrium state depend upon the internal energy (U) and the mass units (mi) of the molecular species i exactly in the same manner as in an equilibrium situation. In other words, each cell of a system in non-equilibrium is treated exactly in the same manner (statistically as well as thermodynamically) as if equilibrium exists in each of these cells". This is termed as the assumption of local equilibrium.

FLUXES AND FORCES

The irreversible thermodynamics is different from the classical thermodynamics in the sense that in the latter when the system is in equilibrium the parameters like temperature, chemical potential, etc. are same at each point of the system whereas in the former; these parameters may increases or decrease from point in the system (say longitudinally *e.g.*, in a metal rod). Thus gradients exist in the system which result in the flow of some kind or the other. For example temperature gradient results in the flow of heat, gradient of chemical potential results in the flow of matter and so on. The gradients such as temperature gradients ; chemical potential gradients etc. is called the driving force whereas the flow of energy, matter etc. (in calories/sec or moles/sec) that takes place due to the gradient is called the *flux*. Thus, the force and flux have a cause-effect relationship *i.e.*, force is the cause and flux is the effect. Similarly, when a potential difference exists at the end of a conductor, a current flows so the that potential difference is the force and flow of current is the flux. Again a reaction takes place when Gibb's free energy change (ΔG) is negative. Thus Gibb's free energy charge is the driving force and rate of flow of reaction (or also called the 'affinity' of the reaction) is called the flux of the reaction.

In general, force is represented by X and the flux by J. The two are related to each other by the relation

$$J = LX \qquad ...(1)$$

where L is the coefficient not depending upon X or J. These coefficients in different relationships are known as phenomenological coefficients and the equations are known as phenomenological equations of thermodynamics.

The equation as given above is the generalisation of the single-flux equations of motion *i.e.*, involving only one type of flow by a particular type of force.

However, if a number of processes occur simultaneously, each force X_i gives rise to the flux J_i. The force and its corresponding flux are called conjugate parameters. Further, a particular type of flux may take place as a result of two or more forces. For example, flow of heat or flow of electricity is dependent on the gradient of electric potential as well as that of temperature. Thus the heat flux (J_i) which will be due to the force of temperature gradient (represented be X_u) and due to the force of electric potential gradient (represented by X_e) will be given by

$$J_u = L_{uu}X_u + L_{ue}X_e \qquad ...(2)$$

L_{uu} and L_{ue} are the phenomenological coefficients of X_u and X_e. They provide a numerical measure of the extent to which the particular driving force contributes to the flux under consideration. The first subscript attached to L represents the nature of the flux while the second signifies the nature of the contributing force.

Analogous to equation (2), the flux of the flow of electricity (J_e) will be given by

$$J_e = L_{eu}X_u + L_{ee}X_e \qquad ...(3)$$

In general, if n processes occur simultaneously, the associated fluxes J_i are linearly dependent on the conjugate forces X_i. *i.e.*

$$J_i = \sum_{j=1}^{n} L_{ij}X_j \qquad ...(4)$$

where $i = 1, 2, \ldots\ldots n.$

Further, it can be shown that when n forces X_i give rise to n fluxes J_i, the entropy production is

$$\sigma = \sum_i J_i X_i \qquad ...(5)$$

i.e., entropy production in the sum of products of forces and fluxes.

ONSAGAR RELATIONS AND APPLICATIONS

Introduction : The three laws of thermodynamics have been used to describe fully the thermodynamics of reversible processes like the dielectric polarisation of a crystal by an electric field etc. However, these three laws furnish only qualitative explanation for the irreversible processes such as that flow (effect) because of temperature gradient (cause), current flow (effect) because of potential difference (cause), etc.

Further, onsagar derived certain relations which are able to describe the quantitative aspects of such irreversible processes. Thus. Onsagar relations are applied to study the thermodynamics of irreversible process.

Onsagar Relations

The thermodynamics of irreversible processes has been based on the following two concepts :

1. Linear Law

This law holds only when system under study is close to equilibrium *i.e.*, only removed slightly from equilibrium. Some examples of this law are Ohm's law between electric current and potential gradient, Fourier's law between heat flow and temperature gradient etc. The causes which give rise to irreversible process)*e.g.*, temperature gradient, potential difference, etc.) are termed as driving forces (denoted by X) and the effects produced (*e.g.*, flow of heat, flow of current, etc.) are flux (denoted by J). Thus, the linear relations, near the equilibrium have been of the following type :

$$J_i = \sum_{k=1}^{n} L_{ik} X_k \quad (i = 1, 2, \ldots n) \qquad \ldots(1)$$

In the above linear relations, the fluxes are proportional to the forces. Relations of this type (Eq. (1)] are termed as phenomenological relations and the coefficients L_{ik} are termed as phenomenological coefficients. The coefficients L_{ik} may denote heat conductivity or electrical conductivity, etc., where are coefficients L_{ik} $(i \neq k)$ may denote the interference of two irreversible processes i and k, e., thermal conduction and diffusion.

2. Onsagar Reciprocal Relations

These are also relations on which thermodynamical of irreversible processes is based. Onsagar gave the following fundamental theorem pertaining to such irreversible processes:

If one makes a 'proper choice' of the fluxes J_i and X_i, the matrix of phenomenological coefficients L_{ik} would be symmetric, i.e,

$$L_{ik} = L_{ik} \ (i_i \ k = 1, 2, \ldots n) \qquad \ldots(2)$$

These identities are termed as Onsagar reciprocal relations. The notion of proper choice of fluxes and forces may be explained as follows.

If the state of the system (local temperature, pressure etc.) may be described by a number of parameters A_1, A_2, ...A_n with their equilibrium values $\overset{\circ}{A}_1, \overset{\circ}{A}_2, \ldots \overset{\circ}{A}_n$, then deviation of state parameters from their equilibrium values is termed as state variable, $\propto$. Hence

$$\propto_i = A_i - \overset{\circ}{A}_1 \qquad (i = 1, 2, \ldots n)$$

Now, we will define the proper choice of fluxes and forces as the time derivative of the state variable $\propto_i$:

$$J_i = \propto_i (i = 1, \ldots n) \qquad \ldots(3)$$

and the following combination of the state variables $\propto_i$:

$$X_i = \frac{\partial}{\partial \alpha_i} (\Delta S) \; (i = 1, 2, \ldots n) \qquad \ldots(4)$$

where, ΔS represents the deviation of the entropy from the equilibrium value.

As entropy gets increased in irreversible process, this change in entropy ΔS from the equilibrium value can be put as a function of the state variables, *i.e.*,

$$\Delta S = \frac{\partial}{\partial \alpha_1} (\Delta S)\, \alpha_i + \frac{\partial (\Delta S)}{\partial \alpha_2} \alpha_2 + \ldots$$

$$= \sum_i \alpha_i \frac{\partial (\Delta S)}{\partial \alpha_2} \qquad \ldots(5)$$

$$= \sum_i J_i \, X_i, \qquad \text{[Using Eqs. (3) and (4)]}$$

In order to illustrate the above derived quantities *e.g.*, fluxes and forces, we will consider a system which is composed of two parts, both enclosed within the same rigid adiabatic enclosure. Suppose the two parts are at uniform temperatures T_1 and T_2. if dQ represents the amount of heat flow from part at temperature T_2 to that at T_1 then increase in enthalpy of the system would be given as follows :

$$dS = dQ \left[\frac{1}{T_1} - \frac{1}{T_2} \right] = dQ. \frac{T_2 - T_1}{T_1 T_2}$$

$$= dQ. \frac{\Delta T}{T^2},$$

where $\quad T_2 - T_1 = \Delta T$

and $T = \sqrt{(T_1 T_2)}$.

Hence rate of entropy production would be as follows :

$$\dot{S} = \dot{Q}\,\frac{\Delta T}{T^2} \qquad ...(6)$$

As heat flow, called flux, denoted by Q occurs due to temperature gradient, called force denoted by ΔT, we can put

$$\dot{Q} = J_i \text{ and } \frac{\Delta T}{T^2} = X_i \qquad ...(7)$$

so that eq. (6) would becomes as follows :

This is thus as eq. (5)

On putting relations (7) in eq. (1) we obtain

$$\dot{Q} = L\,\frac{\Delta T}{T^2}$$

This is nothing but is Newton's law of cooling.

Proof of Onsagar Reciprocal Relations

In order to develop the proof of Onsagar reciprocal relations, we will consider an adiabatically isolated system which is represented by microcanonical ensemble. The developing of the theory of onsagar relations is based on the assumption that the rate at which a macroscopic irreversible process is occurring remains the same as the average rate of decay or regression of the statistical fluctuations in the state of the system. Hence, we shall discuss it in two parts :

Step (i) : Fluctuations of an Isolated System : First of all we will find out the statistical mechanics of fluctuations of a system which remain isolated for a sufficient length to ensure thermodynamic equilibrium. Suppose the state variable is

$$\propto_i = A_i - \overset{o}{A}_i$$

where is same as sated earlier.

The deviation of the entropy from its equilibrium value due to fluctuations $\propto_i$ is denoted by ΔS and is given as follows :

$$\Delta S = -\frac{1}{2}\sum_{i,k=1}^{n}\left(\frac{\partial^2(\Delta S)}{\partial\alpha_i\,\partial\alpha_k}\right)_o \alpha_i\,\alpha_k \qquad ...(1)$$

where, suffix 0 denotes the values of these derivatives at equilibrium. Therms liners in $\propto_i$'s have been excluded as in equilibrium entropy is having maximum value and the state variables are zero. It is possible to further put the above relation in the form

$$\Delta S = -\frac{1}{2}\sum g_{ik}\,\alpha_i\alpha_k \qquad \text{...(2)}$$

Hence $$X_i = \frac{\partial}{\partial \alpha_i}(\Delta S) = -\sum_k g_{ik}\,\alpha_k \qquad \text{...(3)}$$

We know, $$\Delta S = -\sum \alpha_i\, g_{ik}\, \alpha_k$$

$$= \sum_i J_i\, X_i \qquad \text{...(4)}$$

From Einstein's fluctuations theory, it is known that the probability P of fluctuations $d\propto_{1,}$ $d\propto_2$, ... is proportional to exp. (ΔS/k), where k refers to the Boltzmann's constant. Thus, the probability of finding a state in which the values of $\propto_i$ lie between μi, and μi + $d\propto_i$ may be given as follows:

$$P\, d\alpha_1 ... d\alpha_n = \frac{\exp.(\Delta S/k),\ d\alpha_1 ... d\alpha_n}{\int ... \int \exp.(\Delta S/k),\ d\alpha_1 ... d\alpha_n} \qquad \text{...(5)}$$

with normalisation condition given as follows :

$$\int ... \int d\alpha_1 ... d\alpha_n = 1 \qquad \text{...(6)}$$

Now an attempt is made to find the average

$$\overline{\alpha_i\, X_J} \equiv \int ... \int \alpha_i\, X_J\, d\alpha_1 ... d\alpha_n \qquad \text{...(7)}$$

In eqns. (3) and (5) are used, it can be shown that

$$X_J = \frac{\partial(\Delta S)}{\partial \alpha_j} = \frac{\partial}{\partial \alpha_j}(k \log P)$$

$$= k\,\frac{\partial}{\partial \alpha_j}(\log P)$$

If this value is put in Eq. (7), we obtain

$$\overline{\alpha_i\, X_j} = k \int ... \int \alpha_i\, \frac{\partial(\log P)}{\partial \alpha_j}.\ P\, d\alpha_1 ... d\alpha_n$$

$$= k \int ... \int \alpha_i\, \frac{\partial(\log P)}{\partial \alpha_j}.\ P\, d\alpha_j\, d\alpha_1 ... d\alpha_{j-1,}\, d\alpha_{j+i,} ..., d\alpha_n$$

$$= k\int \ldots \int d\alpha_1 \ldots d\alpha_{j-1,}\ d\alpha_{j+1} \ldots d\alpha_n \int \alpha_i \frac{\partial(\log P)}{\partial \alpha_j}.\ P d\alpha_j$$

$$= k\int \ldots \int d\alpha_1 \ldots d\alpha_{j-1,}\ d\alpha_{j+1} \ldots d\alpha_n \int \alpha_i \frac{\partial P}{\partial \alpha_j}.d\alpha_j \quad \ldots(8)$$

By carrying out integration by parts, it can be shown that

$$\int_{-\infty}^{+\infty} \alpha_i \frac{\partial P^{\cdot}}{\partial \alpha_j}\, d\alpha_j = [\alpha_i\, P]_{-\infty}^{+\infty} - \int_{-\infty}^{+\infty} \alpha_i \frac{\partial \alpha_i}{\partial \alpha_j}\, P\, d\alpha_j$$

The first term would be zero as probability of fluctuations would be zero at $\propto_i = \pm\ \infty$. Hence,

$$\int_{-\infty}^{\infty} \alpha_i \frac{\partial P}{\partial \alpha_j}\, d\alpha_j - \int_{-\infty}^{+\infty} \delta_{ij}\ P\ d\alpha_j.$$

where $\delta_{ij} = \dfrac{\partial \alpha_i}{\partial \alpha_j} = 0$ if $i \neq j$

$= 1$ if $i = j$

If this term is substituted in Eq. (8), we obtain

$$\overline{\alpha_i\ X_j} = -\,k \int \ldots \int d\alpha_i \ldots d\alpha_{j-1}, d\alpha_{j+1} \ldots d\alpha_n \int P\, \delta_{ij}\, d\alpha_j$$

$$= -\,k \int \ldots \int P\ d\propto_1 \ldots d\propto_n\ \delta_{ij}.$$

$$= k\ \delta_{ij},\ (i, j = 1, \ldots n) \quad \ldots(9)$$

on using the normalisation condition (6).

In combination with the result of step (ii), Eq. (9) can be used to arrive at Onsagar relation.

Step (ii): This step is involving the concept of microscopic reversibility which means the invariance of equations of motion of individual particles with time reversal. *i.e.*, $t \to -\,t$. Now we will denote these values by $\propto_i(t)$ and $\propto_j(t + \pi)$. The average of their product after a long lapse of time may be put as follows :

$$\overline{\alpha_i(t)\,\alpha_j(t+\pi)} = \lim_{\pi \to \infty \pi} \frac{1}{\pi} \int_0^{\pi} \alpha_i(t)\,\alpha_j(t+\pi)\, dt \quad \ldots(10)$$

This, for one single system (subensemble) would be equal to the average over a microcanonical ensemble which is obtained by using the probability function eq. (6).

Further the average value of the product $\alpha_i(t)\,\alpha_j(t+\pi)$ is different from the average of product given in eq. (10) merely by the substitution

$t \rightarrow -t$. Thus the microscopic reversibility may be put by the relation.

$$\overline{\alpha_i(t)\,\alpha_j(t+\pi)} = \overline{\alpha_j(t)\,\alpha_i(t+\pi)} \qquad \text{...(11)}$$

On subtracting $\overline{\alpha_i(t)\,\alpha_j(t)}$ from both sides and dividing by π, we obtain

$$\overline{\alpha_i(t)\frac{[\alpha_j(t+\pi) - \alpha_j(t)]}{\pi}} = \overline{\alpha_j(t)\frac{[\alpha_i(t+\pi) - \alpha_i(t)]}{\pi}}$$

This, when $\pi \rightarrow 0$, gives

$$\overline{\alpha_i(t)\,\alpha_j(t)} = \overline{\alpha_j(t)\,\alpha_i(t)} \qquad \text{...(12)}$$

Further, it was assumed by Onsagar that on the average the decay or regression of a fluctuation would follow the linear relations of the form

$$\alpha_i = \sum_k L_{ik}\, X_k$$

This when substituted in eq. (12) yields

$$\overline{\sum_k \alpha_i\, L_{ik}\, X_k} = \overline{\sum_k \alpha_j\, L_{ik}\, X_k}$$

The above expression on rearrangement yields

$$\overline{\sum_k L_{ik}\,\alpha_i\, X_k} = \overline{\sum_k L_{ik}\,\alpha_j\, X_k} \qquad \text{...(13)}$$

On combining the result of the above tow steps by using Eqs. (9) and (13), we get

$$-k\sum_k \delta_{ik}\, L_{jk} = -k\sum_k \delta_{jk}\, L_{ik}$$

But $\quad \delta_{ij} = 0$ for $i \neq j$, we get

$$L_{ji} = L_{ij} \qquad \text{...(14)}$$

This is the Onsagar relation.

Application of Onsagar Relation

We will make an attempt to obtain relations that are used describe the thermodynamics of any irreversible process. First of all we will calculate entropy production ΔS and then we will calculate proper fluxes and forces.

We will assume an adiabatically isolated system which is composed of two vessels I and II maintained at temperatures T and $T + \Delta T$. Now a hole is drilled between the two vessels for combination. This will cause thermal flow and mass with interference between the two.

Suppose V is the volume of two vassals. Now U, M and S donate the energy, mass and entropy respectively corresponding to each of the vessels in the state of thermodynamic equilibrium. Further as U and M obey simple conservation laws, they can be chosen as state variables $\propto_i$.

Now the variation of ΔS_I of entropy I is given as follows:

$$\Delta S_I = \left(\frac{\partial S}{\partial U}\right)_M \Delta U + \left(\frac{\partial S}{\partial M}\right)_U \Delta M + \frac{1}{2}\left\{\frac{\partial^2 S}{\partial U^2}\right\}_M$$

$$(\Delta U)^2 + \frac{\partial^2 S}{\partial U\,\partial M}\Delta U\,\Delta M + \frac{1}{2}\left(\frac{\partial^2 S}{\partial M^2}\right)_U (\Delta M)^2 \quad ..(1)$$

Further it is also possible to write a similar expression for chamber II. From conservation laws, we have

$$(\Delta U)_I = -(\Delta U)_{II} : (\Delta M)_I = -(\Delta M)_{II}$$

The total entropy variation of the whole system may be given as follows.

$$2\Delta S = (\Delta S)_I + (\Delta S)_{II}$$

or

$$\Delta S = \frac{1}{2}\left(\frac{\partial^2 S}{\partial U}\right)_M (\Delta U)^2 + \frac{\partial^2 S}{\partial U\,\partial M}\Delta U\,\Delta M + \frac{1}{2}\left(\frac{\partial^2 S}{\partial M^2}\right)_U (\Delta M)^2 \quad ...(2)$$

On taking the time derivative, we obtain

$$\Delta S = \left\{\frac{\partial^2 S}{\partial U^2}\right\}_M \Delta U\,\Delta U + \frac{\partial^2 S}{\partial U\,\partial M}\Delta U\,\Delta M + \frac{\partial^2 S}{\partial U\,\partial M}\Delta U\,\Delta M + \left(\frac{\partial^2 S}{\partial M^2}\right)\Delta M\,\Delta M$$

$$= \Delta U\left[\left\{\frac{\partial^2 S}{\partial U^2}\right\}_M \partial U + \frac{\partial^2 S}{\partial U\,\partial M}\Delta M\right] + \Delta M\left[\left\{\frac{\partial^2 S}{\partial M^2}\right\}_U \partial M + \frac{\partial^2 S}{\partial U\,\partial M}\Delta U\right] \quad ...(3)$$

We have now arrived at the expression of entropy production. Now we will make an attempt to specify the fluxes and forces As there are two flows, there would be two fluxes J_U and J_M denoting energy flow and mass flow respectively. Similarly there are two forces, X_U and X_M.

As fluxes are the time derivative of state variable $\propto_i$, in this case we have

$$\left.\begin{aligned} J_U &= \Delta U \\ JM &= \Delta M \end{aligned}\right\} \qquad ...(4)$$

Also the corresponding forces are given as follows :

$$\left.\begin{aligned} X_U &= \Delta\left[\frac{\partial S}{\partial U}\right]_M = \left[\frac{\partial^2 S}{\partial U^2}\right]_M \Delta U + \left[\frac{\partial^2 S}{\partial U\,\partial M}\right]\Delta M \\ X_M &= \Delta\left[\frac{\partial S}{\partial M}\right]_U = \left[\frac{\partial^2 S}{\partial M^2}\right]_U \Delta M + \left[\frac{\partial^2 S}{\partial M\,\partial U}\right]\Delta U \end{aligned}\right\} \qquad ...(5)$$

If we substitute Eqs. (4) and (5) in (3), we get

$$\Delta S = J_U X_U + J_M X_M \qquad ...(6)$$

The linear relation may be put as follows.

$$\left.\begin{aligned} J_M &= L_{11} X_M + L_{12} X_U \\ J_u &= L_{21} X_M + L_{22} X_u \end{aligned}\right\} \qquad ...(7)$$

and from Onsagar relation, we have

$$L_{12} = L_{21} \qquad ...(8)$$

Eqs. (5) and (7) have been the general expressions for forces and fluxes in the present case. For a simple system at constant volume, we have

$$T\,dS = dU - \mu\,dM \qquad ...(9)$$

where μ is the chemical potential per unit mass. From this relation, we have

$$\left(\frac{\partial}{\partial U}\right)_M = \frac{1}{T} \text{since } dM = 0 \text{ and } \left(\frac{\partial}{\partial M}\right)_U = -\frac{\mu}{T} \text{since } dU = 0$$

so that $$X_U = \Delta\left(\frac{\partial S}{\partial U}\right)_M = \Delta\frac{1}{T} = -\frac{\Delta T}{T^2} \qquad ...(10)$$

and $$X_M = \Delta\left(\frac{\partial S}{\partial M}\right)_U = \Delta\left(-\frac{\mu}{T}\right) = -\Delta\left(\frac{\mu}{T}\right)$$

or $$= -\left(\frac{\Delta\mu}{T} - \frac{\mu}{T^2}\Delta T\right) \quad ...(11)$$

$\because$ $\mu = u + pv - Ts = h \;\; Ts,$

so that $\Delta m = \Delta u + p\,\Delta v + v\,\Delta p - T\,\Delta s - s\,\Delta T = v\,\Delta p - s\,\Delta T,$

Other changes being zero. Putting it in eq. (11) we obtain

$$X_M = -\left(\frac{v\,\Delta p - s\,\Delta T}{T} - \frac{\mu}{T^2}\Delta T\right) = -v\frac{\Delta p}{T} + \frac{\Delta T}{T}\left(s + \frac{\mu}{T}\right)$$

$$= -v\frac{\Delta p}{T} + \frac{\Delta T}{T}\left(s + \frac{u + pv - Ts}{T}\right) = -v\frac{\Delta p}{T} + \frac{\Delta T}{T}\left(\frac{u + pv}{T}\right)$$

$$= -v\frac{\Delta T}{T} + \frac{h\,\Delta T}{T^2} \quad ...(12)$$

If Eqs. (12) and (10) are substituted in Eq. (7) we obtain

and

$$\left.\begin{aligned} J_M &= -\frac{L_{11}v}{T}\Delta p + \frac{L_{11}h - L_{12}}{T^2}\Delta T \\ J_U &= -\frac{L_{21}v}{T}\Delta p + \frac{L_{21}h - L_{22}}{T^2}\Delta T \end{aligned}\right\} \quad ...(13)$$

In the stationary state mass flow would be zero and we can put J_M = 0 so that we get

$$0 = -\frac{L_{11}v}{T}\Delta p + \frac{L_{11}h - L_{12}}{T^2}\Delta T$$

or $$\left(\frac{\Delta p}{\Delta T}\right)_{JM = 0} = \frac{L_{11}h - L_{12}}{L_{11}vT} \quad ...(14)$$

Eq. (14) gives theorem-molecular pressure difference. It is possible to observe the difference of the two effects, namely thermal flow and mass flow by maintaining a certain pressure difference between the two values and a uniform temperature throughout the system. Then the matter would be flowing (mass flow) from one vessel to another associated with thermal energy flow (that is proportional to the flow of matter). Such an effect is tamed as thermomechanical effect and is given as follows :

$$\left(\frac{J_U}{J_M}\right)_{\Delta = 0} = \frac{L_{21}}{L_{11}} = Q^* \quad ...(15)$$

On substituting Eq. (15) in Eq. (14), we get

$$\left(\frac{\Delta p}{\Delta T}\right)_{JM} = 0 = \frac{h - (L_{21}/L_{11})}{vT} = \frac{h - Q^*}{vT} \qquad ...(16)$$

Eq. (15) connects the two effects.

STATES OF STATIONARY

The quite often we study systems in a state called the 'stationary state' as it can be readily subjected to mathematical analysis in a number of situations of practical importance.

A system is said to be in a stationary state if its macroscopic properties such as temperature, pressure, composition and entropy do not change with time inspite of the possible occurrence of irreversible processes. Of the macroscopic properties, the intensive ones, though unchanged in time, will generally still vary from point to point in the system.

The concept of a stationary state is illustrated by the following examples :

(i) If heat is added at a constant rate to one end of a metal bar and withdrawn at an equal rate from its other end, the temperature at each point of the bar approaches a time independent value. All the same, the temperature varies along the length of the bar and entropy is produced continuously as a result of heat conduction.

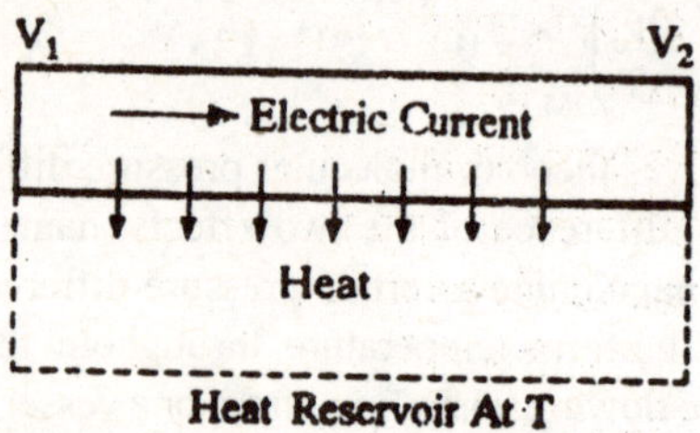

Fig. 3.1

(ii) Consider a metal rod between whose ends there is a potential difference ΔV. An electric current flows though the rod. Suppose that the rod is placed along its length in contract with a heat reservoir at the temperature T, and that the rate of heat flow out of each element into the reservoir is equal to the rare at which the electrical work is done on the element. All the properties of the rod (temperature, internal energy, entropy etc.) then remain Constant with time inspite of the irreversible generation of heat (due to flow of electric current).

The stationary states may be classified into the following two classes :

(i) Equilibrium stationary states *i.e.*, the stationary states attained in a system in equilibrium.

(ii) Non-equilibrium stationary states *i.e.*, the stationary states attained in a system not in equilibrium.

The two differ mainly in the fact that in the former, the value of any intensive property not only remains constant with time but also has the same value at every point whereas in the latter, the value of any intensive property though remains constant with time at any point but differs from point to point and thus a continuous inflow of energy or matter and energy is required from the surroundings to the system. Thus, whereas equilibrium stationary states can be attained only in an isolated system (or a system in contact with a uniform environment), the non-equilibrium stationary states can be attained in open systems.

Another method of classifying stationary states is in the following manner :

Consider a system characterised by n independent forces X_1, X_2..., X_3, and let a number k of them, say X_1, X_2..., X_k, be kept at fixed values through the operation of external constraints. E is then found empirically that the system will sooner or later reach a stage where the remaining forces, X_{k+i}, X_{k+2}...X_n, will also remain constant with the passage of time. In the end, therefore, all the forces X_i. and as a result, all the fluxes J_i (depending linearly on the forces) become independent of time. The stationary state attained by this method is called the *stationary state of order k.*

In the special case, when no forces X_i are held fixed (k = 0). The system will continue till it reaches a stage when all the fluxes, and consequently all the forces, have vanished. Thus, the stationary state of zero order is identical with the state of thermostatic equilibrium. The theory of stationary states related to open system has applications of considerable importance to living phenomena (biological systems). It has been suggested that if the living organisms and cells are treated as open systems, the basic biological phenomena.

INTRODUCTION TO CLASSICAL (STATISTICAL) THERMODYNAMICS

The classical thermodynamics is based on empirical laws in which the variables such as pressure, temperature and energy are regarded as functions of state. However, in the derivation of relations of thermodynamic functions, we ignore the constitution of the material system. But the experimental observations support the laws. The only drawback in classical thermodynamics is that we cannot calculate directly the absolute value off any thermodynamics function of a substance. In order to over come this, a new method of approach has been developed and is known as statistical thermodynamics. The statement according, to which non-spontaneous process also called negative process cannot be the only result of a complex of processes is not strict are negative processes in microscopic (bulk) system events. Therefore the second law of thermodynamics first law, but a statistical law that is observed with a high degree off accuracy for considerable numbers of molecules and is less applicable, the smaller are the dimensions of the system being studied. So the second law of thermodynamics has statistical nature. This fact was suggested by L. Bottzmann. He explained in this way the contradiction between the reversibility of mechanical motion and the irreversibility and unidirectional nature of real physical and chemical processes; this unidirectional nature is the result of the molecular structure of the material world. In this way, it was Boltzmann who provided the molecular basis to chemical thermodynamics. This approach is called chemical statistics or statistical thermodynamics and it deals with the applications of mechanics to chemical thermodynamics.

Statistical thermodynamics has bee applied successfully to relate the microscopic properties of the individual molecules (*e.g.* moment of inertia, dipole moment etc.) with macroscopic properties (*e.g.* molar heat capacity, polarization, etc.) of a system having a large number of molecules. In supposed to be composed of innumerable particles in motion. This motion of particles is governed by the laws of dynamics and in fact subject to quantum the number of particles is so large that it is impossible to follow the behaviour of every individual particle Statistical methods can be used to find out the properties of the system, particularly to know what fraction of the molecules has a given energy. It is possible to establish that the laws of thermodynamics are the necessary consequence of the quantum principles, when statistical postulates are used. In this method, we take into account the atomic constitution of matter. The matter is supposed to consist of innumerable particles in motion is governed by the laws of dynamics.

Statistical method can be employed to express the property o the system to find out what fraction of the molecules has a given energy. It is easier to find out the values of entropy, free energy, enthalpy, heat capacity etc., from spectroscopic data by using statistical methods. We shall however, discuss the equilibrium aspects of the system in the succeeding articles. In other words, this is a new approach that deals with the applications of statistical mechanics to thermodynamics. It has been applied successfully to relate the microscopic properties of the individual molecules (moment of inertia, dipole moment, etc.) with macroscopic properties (molar heat capacity, polarization, etc.) of a system having a large number of molecules. Statistical thermodynamics can be used for systems at equilibrium as well as away from equilibrium. We shall limit our discussion to the equilibrium aspects of the system Distinction between quantum mechanics, statistical mechanics and statistical thermodynamics. Statistical mechanics acts as a bridge between thermodynamics and quantum mechanics.

Quantum mechanics provides information about the energy of the molecular system, statistical mechanics tells us about the possible arrangement of the energy among various molecules of the system and introduces the concept of probability and partition functions. Statistical thermodynamics deals, with the relationships between the probability partition functions and thermodynamic properties. In other words, a physical entity generally manifests itself in : (i) A Microscopic, and (ii) A Macroscopic state. The dealt with macroscopic or the bulk state of matter in the preceding chapters.

The macroscopic variables were pressure, volume, temperature, entropy, etc. The macroscopic variables were related mutually on thermodynamical considerations. Further, we assumed that the number of atoms/molecules in them (microscopic details of a macrosystem) were exorbitantly large $\approx 10^{25}$ m^{-3}. The microscopic systems, on the other hand, have very few particles. Now, we know that the actual systems, the macro systems, are formed of the microsystems and so there should be a definite relation between the two.

Classically, if we want to have a complete description of a system we need to know actual velocity (or momentum), location and mutual interaction (if any, with other particles) of every particle. Since every panicle is described by three positions and three momentum coordinates, hence to describe a macrosystem we need a very large number of data. Further, any change in the parameter of an individual particle should be reflected in the properties of the system as a whole. *This poses a question*: How to deal with such a vast data which even a super-computer is not capable to handle, if at all we are able to collect it. If, however, we remember that:

(i) Much of the data and analytical results obtained on micro level are not useful in the study of the macrosystem, and

(ii) The application of classical dynamical principles to a many particle system is theoretically inapplicable, we can easily dispense with the technically impossible and practically useless classical kinetic and thermodynamic methods in favour of new statistical methods.

In the following sections we will see the importance of the concept of probability in the development of the statistical theory. Hence, we shall first consider some elementary theorems on the calculus of probability. In other words, it refers to a class of events of which the particular event is one member. Mathematically, the probability is defined as: When a certain aspect is possible in a ways and is not possible in b ways, then the probability of occurring the aspect is a/(a + b). (Newton defined probability as the ratio of a certain number of aspects, which were possibly observable, to the total number of aspects included in that class.) Here (a + b) is the total number of aspects each of which has the same chance of occurrence. Statistics is applied only to such systems which contain a large number of identical individuals which may be distinguishable by means of some property. In a statistical method, we study the behaviour of aggregates of particles to calculate the average value without investigating the detailed behaviour of the system and finding there from the value for each individual. Statistically, it is sufficient to know the most probable distribution of particles in different accessible states. This branch of physics has seen its development in the hands of Maxwell, Boltzmann, Gibbs, Fermi-Dirac, Bose-Einstein etc.

Now, as we have shown, if an event can occur in a ways, the probability of occurring of this event is a/(a + b), and similarly, if it fails to occur in b ways, then the probability of its failing is b/(a + b). The sum of the two

$$\frac{a}{a+b} + \frac{b}{a+b} = 1$$

corresponds to the certainty of some event, since the event must occur or fail to occur. If b = 0, the probability of occurrence becomes certain but if a = 0, the probability of its failure is certain.

We may define the probability for a certain event occurring m times in n unrestricted (free) attempts as,

$$P = \lim_{n \to \infty} \frac{m}{n}.$$

THEOREMS OF PROBABILITY IN STATISTICAL THERMODYNAMICS

The following are the important probability theorems commonly used in statistical thermodynamics:

(1) The number of ways in which N distinguishable particles can he arranged in order is equal to

$$N!$$

(2) The number of different ways in which N particles can be selected from N distinguishable particles irrespective of the order of selection is equal to

$$\frac{N!}{(N-n)!n!}$$

(3) The number of different ways in which n indistinguishable particles can be arranged in g distinguishable states with not more then one particle in each state is equal to

$$\frac{g!}{(g-n)!n!}$$

PROBABILITIES CALCULATION

We can extend the theory of probability in order to include the occurrence of two or more independent events. The combined probability in such cases is given by the product of separate probability of each event. Let us examine some cases:

(i) *Tossing of a coin:* On tossing a single coin, there are two possible aspects that may be presented by a fallen coin, a "head" or a "tail". We may assume here that there is no preference for any particular event, so that after making a large number of throws, the ratio of the number of heads to the number of tails is unity. Thus, for a normal coin the probability of each event is 1/2.

(ii) *Tossing of two unlike coins*: Suppose we toss two coins of different denominations or of different sizes, or of different colours, we have four complexions.

1. $Head_1$ $Head_2$
2. $Head_1$ $Tail_2$
3. $Tail_1$, $Tail_2$
4. $Tail_1$ $Head_2$.

Thus, the probability of any one complexion on a single toss will be 1/4 = 1/2.1/2, Or the composite probability is the product of probability of individual events.

(iii) *Tossing of two similar coins*: If two coins cannot be distinguished from each other, then complexions 2 and 4 of (ii) are indistinguishable whence the probability of one head and one tail is twice than that of two heads or two tails.

(iv) *Tossing of two coins in general*: Let us represent occurrence of heads by A_1, A_2 and the occurrence of tails by B_1, B_2 when two coins are tossed. Then the composite event can be represented by $(A_1 + B_1)(A_2 + B_2) = A_1A_1 + A_1B_2 + A_2B_1 + B_1B_2$.

(v) *Tossing of three unlike coins:* With three unlike coins, we have three distinguishable heads A_1, A_2, A_3 and three distinguishable tails, B_1, B_2, B_3. Here we can have eight equally probable complexions $A_1 A_2 A_3$. $A_1 A_2 B_3$, $A_1 B_2 B_3$, $B_1 A_2 B_3$, $B_1 A_2 B_3$, $B_1 B_2 A_3$ and $B_1 B_2 B_3$ which are really obtained from the product of $(A_1 + B_1)$, $(A_2 + B_2)$, $(A_3 + B_3)$.

(vi) *Tossing of three similar coins:* In this case we have only four probable states with complexions

1. All heads.
2. Two heads and one tail
3. All tails.
4. One head and two tails.

We can tabulate the result as in (iv) for

$(A + B)(A + B)(A + B) = A_3 + 3A^2B + 3AB^2 + B^3$.

(vii) *Tossing of n coins*: Suppose all the coins are different. Then the possible complexions (also known as probability number) will be $(A_1 + B_1)(A_2 + B_2) \ldots (A_n + B_n)$. And, if the coins are all alike the total number of possible complexions will be

$$(A + B)^n = A^n + {}^nC_1 A^{n-1} B + {}^nC_2 A^{n-2} B^2 + \ldots {}^nC_r A^{n-r} B^r + \ldots + {}^nC_n B^n$$

$$\text{or} \qquad = \sum_{r=0}^{r=n} C_r A^{n-r} B^r$$

$$\text{or, also } (A + B)^n = \sum_{r=0}^{r=n} C_r A^r B^{n-r}$$

$$\equiv \sum_{r=0}^{r=n} C_r A^{n-r} B^r$$

where $${}^nC_r = \frac{n!}{r!(n-r)!} \equiv \text{Number of events with r heads/tails.}$$

Now, since A and B indicate one event each, the total number of events are $(1 + 1)^n = 2^n = {}^nC_1 + {}^nC_2 + \ldots + {}^nC_n$

$\therefore$ The probability of r heads and $(n - r)$ tails to appear is

$$P(r, n-r) = \frac{{}^nC_r}{2^n} = \frac{n!}{r!(n-r)!} \cdot \frac{1}{2^n}$$

Maximum and minimum probability: To find out which of the combination of heads and tails is most likely in a large number of throws of n coins, we have to find the maximum value of ${}^nC_r$. It is known from algebra that the value of ${}^nC_r$ is greatest for $r = n/2$ when n is even and $r = n+1/2$ when n is odd.

$$P_{max} = \frac{{}^nC_{n/2}}{2^n} = \frac{n!}{(n/2)!(n/2)} \cdot \frac{1}{2^n}$$

Similarly, the minimum probability, P_{min} obtained for $r = 0$ or $r = n$ as ${}^nC_n = 1$.

whence $\qquad P_{min} = 1/2^n$

(viii) *Tossing of a single coin n times*: When we toss the same coin second, third,... nth time, its behaviour is again unpredictable. It may show a head or a tail each time with 1/2 probability. Hence a single coin tossed n times (and of course, making large number of such attempts, if we want to have satisfactory results) is just equivalent to the case of tossing of n coins as discussed in (vii).

Another equivalent illustration can be: Take 2 boxes, identical in all respects, and throw n balls from a great distance. Assuming that each ball will fall into one or the other box, the distribution of balls in the boxes will be governed by the relation (vii),

(ix) *Tossing of a dice*: If we toss of dice, having six faces, the probability for each face showing up will be 1/6. Since now the total number of aspects is 6 and only one can show up at a time with equal probability. Again, we may take 2, 3,.... n dices and throw them and calculate the composite probability for a certain complexion.

(x) *Tossing of n dices each having m faces*: Here let us assume that dice 1 has faces a_1, a_2, a_3... a_m; dice 2 has faces b_1 b_2 b_3 ... b_m; dice 3 has faces c_1, c_2, c_3 ; etc., then the possible complexions with n dices will be the sum of each aspect

a_1, a_2, a_3... a_m, v_1, b_2, b_3 ... b_m , c_1, c_2, c_3 ... n sets and, if the dice are all alike, we have

$m_{.a}$, $m_{.b}$, $m_{.c}$... n sets.

If the probability for occurring each of the face is same, *i.e.*, a = b = c... = one aspect, we get total number of possible aspects

= m.m.m.... n times = m^n

One can show that the most probable distribution corresponds to equitable distribution of faces *i.e.*, face 1, face 2, face 3, etc., will appear in equal number. Then all the dices showing face 1 have the probability of only $1/m^n$.

A similar illustration to this is : Take *m identical boxes and throw n* balls onto them from a great distance. We will take up this problem later in more detail. As in actual analysis of physical problems we have a large number of particles which are distributed into a large number of (hypothetical) boxes.

(xi) *Probability with weightage*: So far we have considered the case in which the probability of occurring an event and its failure is the same. But there can be cases where the probability of occurring a particular event is greater than that for any other event.

Let a be the probability of occurring an event and b be the probability of its failure, per trial. Then, if the trial is repeated n times, the probability of r favourable events is

$$a^r.b^{n-r} = a^r (1 - a)^{n-r}$$

But there are ${}^nC_r$ ways for r favourable events, therefore, the probability of occurring r events is

$$P(r, n - r) = {}^nC_r\, a^r (1 - a)^{n-r}$$

Thus, for example, in a dice a = 1/6

$$\therefore \quad P(r, n - r) = {}^nC_r \left(\frac{1}{6}\right)^r \left(\frac{5}{6}\right)^{n-r}$$

But, if certain number, say 3, has a = 2/7 then in n tosses the number 3 will occur r times with a probability.

(xii) *Postulate of equal a priori probability:* It states that all accessible microstates corresponding to a macrostate are equally probable. Thus, the postulate claims that the probability of finding a molecule in one region of phase space is identical with that for any other region of equal volume provided the regions are obeying other conditions of equality (*i.e.*, the regions have extensions of same magnitude and correspond to the same energy). Thus, it follows that the probability of occurrence of a macrostate is proportional to the number of microstates associated with it

i.e., $P_i \propto \Omega_i$

where P_i is the probability and Ω_i is the total number of accessible microstates, in the ith set.

STATISTICAL POSTULATES

Following postulates are made for applying the statistical method to physical systems:

(i) Every system has a large number of particles/molecules which are in constant random motion.

(ii) The motion of particles/molecules is described in 2Nf-dimensional phase-space.

(iii) The phase-space is divided into cells.

(iv) Postulate of equal a priori probability holds good.

(v) Probability of a particular configuration is given by the ratio of microstates corresponding to that configuration to total accessible microstates, *i.e.*,

$P_i = \Omega_i/\Sigma\Omega_i$ and the average value of any physical parameter is given by

$$<x> = \Sigma x_i P_i = \frac{\Sigma x_i \Omega_i}{\Sigma \Omega_i}$$

where Qi is the frequency of the configuration.

(vi) The equilibrium state of the system is the state of maximum probability.

(vii) The law of conservation of mass is generally valid whence

$$n_1 + n_2 + n_3 + ... = \Sigma n_i, = N = \text{constant.}$$

(viii) The law of conservation of energy governs the distribution of particles among cells.

Thus, if ε_1, ε_2, ε_3 ... denotes energies of each particle in cell 1, 2, 3, etc., then

$$E = \varepsilon_1 n_1, + \varepsilon_2 n_2 + \varepsilon_3 n_3 + ... = \Sigma\varepsilon_i n_i = \text{constant.}$$

(ix) Every point in the phase-space (cell) represents a microstate, the cell volume being h where/is the degree of freedom.

(x) The number of microstates in the energy interval e and $\varepsilon + d\varepsilon$, per unit volume is represented as $\Omega(\varepsilon)$.

(xi) A collection of accessible microstates form a statistical ensemble.

ENSEMBLE METHOD

Ensemble method is superior to the method of *time average*, in determining average value of a physical quantity. So first we recall the characteristics of an ensemble and its way of representation.

1. It is a group of large number of systems.
2. Every member of an ensemble is an exact miniature copy of the system under test.
3. Every member of an ensemble is subjected to the same experimental conditions.
4. There may be three kinds of interactions between the members of the ensemble:

(a) Thermal-related to T,

(b) Mechanical-related to V, and

(c) Diffusive-related N or m (*i.e.*, chemical potential).

5. Every member of the ensemble may have a large number of microstates (also called the dynamic state) due to random motion of its particles/molecules/ And when the configuration of the member systems is maintained same in terms of generalised coordinates and momenta, the statistical method is expected to yield quite accurate results.

Ensemble Representation

Ensembles can be represented in a Γ-space by the distribution of its phase-points. Generally, the distribution of phase-points in an ensemble is continuous and can be represented by a density function ρ(p, qt).

The number of phase-points in the vicinity of (p, q) in Γ-space is given by p (p, q t) dT where d r = dp dq = volume element of Γ- space.

If ρ (p, q t} is known at time t_1, its value can easily be obtained for some other time t_2. Now, from the probability theory

$$d\rho\ (p, q)\ \alpha\ d\Gamma \qquad \therefore d\rho\ (p, q) = \rho(p, q)d\Gamma$$

Here ρ (p, q) is known as the distribution function. The ensemble average of any function R is then given by

$$\langle R\rangle = \int R\ (p, q)\ \rho(p, q)d\Gamma.$$

Selection of Ensemble

(i) Molecules of real gases are usually formed of 2, 3.... atoms whence the density of phase points is not uniform,

(ii) The density of phase points is non-zero for every range of energy,

(iii) Total number of particles and energy is conserved. Point (i) eliminates use of uniform ensemble. Point (ii) eliminates use of microcanonical ensemble. Point (in) eliminates use of grand canonical ensemble to real gases.

But since the properties of these gases such as N, V, T, ε, H etc., are constant the canonical ensemble finds use in description of real gases for their thermodynamic properties.

ACCESSIBLE MICROSTATES

Let us now count the number of energy states available in a macrosystem. Let E be the total energy of this system which has been divided into small energy intervals δ E so that a sufficient number of quantum stales are available in each interval of δ E. Now, if Ω (E) is the number density of such states in the energy interval E and E + δE then

$$\Omega(E) = \rho(E)\ \delta E \qquad ...(i)$$

where ρ (E) is density of states (the number of states in unit energy interval). Thus

$$\Omega(E) \propto \delta E \qquad ...(ii)$$

Let the value of Ω(E) changes with the magnitude of energy interval. Hence we may consider Ω(E) as a smoothly varying function of E.

Again, if ϕ(E) denotes the number of states with energy up to E and ϕ(E + δE) denotes the number of states with energy up to E + δE then

$$\Omega(E) = \phi(E + \delta E) - \phi(E) \equiv \frac{\delta\phi}{\delta E}\delta E \qquad ...(iii)$$

Comment on ρ (E). Relations (i) and (ii) (possibly) indicate that ρ(E) is a constant. But it is not so. In fact ρ (E) is an increasing function of E. Let us see how! We know that the total number of microstates is equal to the product of thermodynamic probabilities of different macrostates (composite probability is obtained, always by multiplying probabilities for different complexions), *i.e.*,

$$\Omega = \Omega_1\, \Omega_2\, \Omega_3 \ldots \Omega_i \ldots \Omega_f$$

or

$$\Omega = \prod\nolimits_1 \Omega_i \ (i = 1 \text{ to } f)$$

where f is the degree of freedom.

But Ω_i should be somehow related to E, *i.e.*,

$$\Omega \propto E^a \quad \therefore \Omega \propto E^{fa} \qquad ...(iv)$$

usually a ≅ 1/2 and/ranges between 10^{22} and 10^{28}. This suggests that for a certain energy interval δE, ρ(E) increases with the value of E.

QUANTUM STATISTICS

Maxwell-Boltzmann distribution, it has been assumed that all the energy levels are accessible to all the particles of the system. However, there may be certain levels prohibited to a certain group of particles. Each energy state is associated with a certain available wave function.

The probability of a particular distribution is restricted by the available wave functions of any state. These restrictions are taken into account in quantum statistics. There are two types of quantum statistics

(i) Fermi-Dirac statistics

(ii) Bose-Einstein statistics.

In Fermi-Dirac statistics the particles are assumed to obey Pauli's exclusion principle. They are characterised by antisymmetric wave functions. The particles under, this category are called *fermions*. Protons neutrons and electrons are fermions.

In the case of Bose-Einstein statistics, it is assumed that the particles are not restricted by Pauli's exclusion principle and are characterised by symmetric wave functions. These particles are called *bosons*. It has been experimentally found that all particles having spin zero or integral multiple of 1 are bosons. Helium nuclei and mesons are bosons.

In both the kinds of quantum statistics, the particles are identical and indistinguishable. At high temperature and low pressure, all the three statistics give practically the same result.

STATISTICAL MECHANICS

Statistical Mechanics deals with systems consisting of many particles and the methods employed are to get a collective or macroscopic property of the system without taking into account the individual motion of the particles The word particle is used in a broad sense meaning a fundamental particle *e.g.*, an electron, an atom, a molecule etc. The particle is a well defined and a stable unit of a given physical system. As it is not practically possible to determine the property of each particle individually, a statistical approach is made by using the concept of probability of distribution. This statistical approach helps in determining the bulk or macroscopic property of the system as a whole. The idea of probability does not imply that the particles move in a random way without obeying the laws. For a many particle system, the statistical analysis is valid regarding the applications of probability distribution of particles.

STATISTICAL EQUILIBRIUM

Consider that a thermodynamical isolated system consists of N particles. The energy states available to the particles are E_1, E_2, E_3 etc. These energy states may be quantized or may be continuous and are due to vibrational and rotational energy of the particles.

Suppose, that at any given instant of time, n_1 particles are in state of energy E_1, n_2, particles with energy state E_2 and so on.

The total number of particles in the system is,

$$N = n_1 + n_2 + n_3 + ... = \sum_i n_i \quad ...(i)$$

Here $i = 1, 2, 3$ etc.

The total energy of the system,

$$U = n_1E_1 + n_2E_2 + n_3E_3 + ...$$

$$U = \sum_i n_i E_i \quad ...(ii)$$

Equation (ii) refers to the total energy of a system in which the particles are non-interacting. Here, the energy of each particle depends only on the coordinates of the particle in the system. For an isolated system, the total energy U is constant.

For isolated system.

$$U = \sum_i n_i E_i = \text{constant} \quad ...(iii)$$

Consider a gas having N molecules at a certain temperature and pressure. Its volume, temperature and pressure are kept constant *i.e.*, the system is isolated. The total energy of this system remains constant. But, the molecules of the gas collide with each other and also with the walls of the container. Consequently, the number of molecules change from one energy state to the other energy state. It means that the value of n_1, n_2, n_3 etc., continuously change. It can be reasonably assumed that for each microscopic State of a system of particles, there is a particular most favoured distribution. When this distribution or partition is reached, the system attains statistical equilibrium. For an isolated system, the values of n_1, n_2, n_3 etc., vary only near the values corresponding to the most probable distribution. Hence, the basic problem in statistical thermodynamics is to obtain the most probable distribution law for a given composition of the system.

In practice, three most probable distribution laws are used. They are:

(i) Maxwell-Boltzmann Distribution Law.

(ii) Fermi-Dirac Distribution Law (Statistics).

(iii) Bose-Einstein Distribution Law (Statistics).

MAXWELL-BOLTZMANN DISTRIBUTION LAW

Consider a system that contains a large number of particles that are identical and distinguishable. The identical particles refer to the particles having the same structure. The distinction between the particles is due to their energy states at a given instant. Four particles are in energy state E_1, 2 in state E_2, zero in state E_3 and so on. It is assumed that all the energy states are accessible to each particle. Consequently, it can be assumed that the probability of any particular partition is proportional to the number of different ways in which the particles can be distributed in the existing available energy states so as to produce the desired partition.

To obtain the required distribution as the first particle a in state E_1 can be selected in N ways. The second particle b in state E_1 can be selected in (N – 1) ways and so on. Therefore, the total number of ways in which the first four particles in state E_1 can be selected is given by,

$$N(N-1)(N-2)(N-3) = \frac{N!}{(N-4)!}$$

Moreover, the four particles in state E_1 can be arranged in 4! different orders. For example, *abcd, bcda, cdab* and so on. There are 24 ways. But, it is immaterial for these particles to be arranged in any particular order in state E_1, because they are identical. Thus, the total number of *distinguishable* different ways are,

$$\frac{N!}{4!(N-4)!}$$

In general, if the first state consists of n_1 particles, the distinguishable different ways, for arranging n_1 particles in state E_1, are,

$$p_1 = \frac{N!}{n_1!(N-n_1)!} \qquad \text{...(i)}$$

For the second state E_2, only (N – n_1) particles are available and n_2 particles are in state E_2. The number of distinguishable different ways are,

$$p_2 = \frac{(N-n_1)!}{n_2!(N-n_1-n_2)!} \qquad \text{...(ii)}$$

If this process is continued for all the available states, the total number of distinguishable ways are obtained by multiplying P_1, P_2, P_3 etc.

$$P = p_1 \times p_2 \times p_3 \times \ldots\ldots$$

$$P = \left[\frac{N!}{n_1!(N-n_1)!}\right]\left[\frac{(N-n_1)!}{n_2!(N-n_1-n_2)}\right]\times\ldots.$$

$$P = \frac{N!}{n_1!n_2!n_3!\ldots} \qquad \ldots\text{(iii)}$$

The distinguishable ways are,

$$P = \frac{N!}{4!2!0!3!1!} \qquad \ldots\text{(iv)}$$

Here 0! is equal to one.

It has been assumed so far that all the available states have the same probability of occupation by the particles. However, it may happen that the states have different intrinsic probabilities say g_i.

For example, a particular energy state may be favourable with more different angular momentum states than the rest and hence it is more likely to be occupied. Taking into account this intrinsic probability factor, the value of P will be different.

If g_i is the probability of locating a particle in a certain energy state E_i, then the probability of locating 2 particles in the same state is $g_i \times g_i = g^2_i$. For n_i particles, the probability is $g_i n_i$. Hence the total probability for a given distribution is given by

$$P = \frac{N! g_1 n_1 g_2 n_2 g_3 n_3 \ldots}{n_1!n_2!n_3!n_4!\ldots} \qquad \ldots\text{(v)}$$

Here n_1, n_2, n_3 etc., are the number of particles in states E_1, E_2, E_3, etc. and g_1, g_2, g_3 etc., are the intrinsic probabilities for states E_1, E_2, E_3 etc.

If all the particles are further assumed to be indistinguishable *i.e.*, particles in state E, and particles in state E_4 cannot be distinguished, then N! permutations among the particles themselves and occupying the different states result in the same distribution. The probability in this case is given by

$$P = \frac{1}{N!}\left[\frac{N! g_1 n_1 g_2 n_2 g_3 n_3 \ldots}{n_1!n_2!n_3!\ldots}\right]$$

$$\therefore \quad P = \frac{g_1 n_1 g_2 n_2 g_3 n_3 \ldots}{n_1! n_2! n_3! \ldots}$$

$$\therefore \quad P = \prod_{i}^{N} \frac{g_i n_i}{n_1!} \qquad \ldots(vi)$$

Here Π is the product sign (since probability is the product of such distributions.

The most probable distribution can be obtained by evaluating the maximum value of $\log_e$, P in equation (vi). This should also satisfy the two conditions that

$$\sum_i n_i = N \qquad \ldots(vii)$$

$$\text{and} \quad \sum_i n_i \; E_i = U \qquad \ldots(vii)$$

According to Stirling's approximation

$$\log_e x! = x \log_e x - x$$

From equation (vi), applying Stirling's approximation, we get

$$\log_e P = \sum_i (n_i \log_e g_i - \log_e n_i!)$$

$$= \sum_i [n_i \log_e g_i - (n_i \log_e n_i - n_i)]$$

$$= \sum_i (n_i \log_e g_i - n_i \log_e n_i + n_i)$$

$$= \sum_i n_i - \sum_i (n_i \log_e n_i - n_i \log_e g_i)$$

$$\log_e P = N \sum_i n_i \log_e \frac{n_i}{g_i} \qquad \ldots(ix)$$

Differentiating equation (ix)

$$d(\log_e P) = \sum_i (dn_i) \log_e \frac{n_i}{g_i} - \sum_i n_i d\left(\log_e \frac{n_i}{g_i}\right)$$

$$= \sum_i dn_i \log_e \frac{n_i}{g_i} - \sum_i n_i \frac{dn_i}{n_i}$$

$$= \sum_i dn_i \log_e \frac{n_i}{g_i} - \sum_i dn_i$$

But $\sum_i dn_i = 0.$

$$\therefore \quad -d(\log_e P) = \sum_i \left[\log_e \left(\frac{n_i}{g_i} \right) \right] dn_i \qquad \text{...(x)}$$

To obtain the maximum value of P,

$$d(\log_e P) = 0$$

$$\therefore \quad \sum_i \log_e \left(\frac{n_i}{g_i} \right) dn_i = 0 \qquad \text{...(xi)}$$

But $\quad \Sigma dn_i = 0 \qquad$ ...(xii)

and $\quad \Sigma E_i dn_i = 0 \qquad$ ...(xiii)

Multiplying (xii) by α and (xiii) by β and adding to equation (xi), we get

$$\sum_i \left[\log_e \left(\frac{n_i}{g_i} \right) + \alpha + \beta E_i \right] dn_i = 0$$

The equilibrium distribution is possible if

$$\log_e \frac{n_i}{g_i} + \alpha + \beta\, Ei = 0 \qquad \text{...(xiv)}$$

or $$\frac{n_i}{g_i} = e^{-\alpha-\beta Ei}$$

or $$n_i = g_i\, e^{-\alpha-\beta Ei} \qquad \text{...(xv)}$$

This gives the maximum probability distribution and a and p are two parameters that depend upon the physical property of the system.

$$N = n_1 + n_2 + n_3 + \ldots$$

$$N = g_1 e^{-\alpha-\beta E1} + g_1 e^{-\alpha-\beta E2} + \ldots.$$

$$N = e^{-\alpha}[g_1 e^{-\beta E1} + g_2 e^{-\beta E2} + \ldots.]$$

$$N = e^{-\alpha}[\sum_i g_i\, e^{-\beta Ei}]$$

Take $[\sum_i g_i\, e^{-\beta Ei} = Z \qquad$ ...(xvi)

Here, Z is called the partition function.

$$N = e^{-\alpha}(Z) \qquad \text{....(xvii)}$$

$$e^{-\alpha} = \frac{N}{Z}$$

Substituting this value in equation (vii)

$$n_i = \left(\frac{N}{Z}\right)[g_i e^{-\beta Ei}] \qquad \text{....(xviii)}$$

This equation refers to Maxwell-Boltzmann Distribution Law.

MAXWELL-BOLTZMANN DISTRIBUTION IN TERMS OF TEMPERATURE

The total energy of an isolated system is given by

$$U = \sum_i n_i E_i$$

$$U = n_1E_1 + n_2E_2 + n_3E_3 + ...$$

$$U = (g_1e^{-\alpha-\beta E1}) E_1 + (g_2e^{-\alpha-\beta E2}) E_2 + ...$$

$$U = e^{-\alpha}[g_1E_1e^{-\beta E1} + g_2E_2e^{-\beta E2} + ...]$$

But $e^{-\alpha} = \dfrac{N}{Z}$

$$\therefore \quad U = \frac{N}{Z}[g_1E_1e^{-\beta E1} + g_2E_2e^{-\beta E2} + ...]$$

$$U = \frac{N}{Z}\sum_i g_i E_i e^{-\beta Ei} \qquad \text{...(i)}$$

Here $Z = \sum_i g_i e^{-\beta Ei}$ ...(i)

$$\frac{dZ}{d\beta} = \frac{d}{d\beta}\sum_i g_i e^{-\beta Ei} = -\sum_i g_i E_i e^{-\beta Ei}$$

$$\therefore \quad \sum_i g_i E_i e^{-\beta Ei} = \frac{d}{d\beta}\sum_i g_i E_i e^{-\beta Ei}$$

Substituting this value in equation (i)

$$U = -\frac{N}{Z}\frac{d}{d\beta}\sum_i g_i e^{-\beta Ei}$$

$$U = -\frac{N}{Z}\frac{d}{d\beta}$$

$$U = -N\frac{d}{d\beta}[\log_e Z] \qquad \text{...(ii)}$$

The average energy of a particle is

$$E_{av} = -\frac{U}{N} = -\frac{d}{d\beta}(\log_e Z) \qquad \text{...(iii)}$$

This shows that for a given system, the total energy U, the partition function Z, and the average energy of the particle E_{av}, depend on the parameter β.

Therefore, β may be taken to characterise the internal energy of the system and has the units per joule. If T is the temperature in degrees kelvin, then it is more convenient to represent parameter β as

$$\beta = \frac{1}{kT} \text{ or } KT = \frac{1}{\beta} \quad ...(iv)$$

Here kT has the unit of energy *i.e.*, joule and k is the Boltzmann's constant. Its units are J/K. The value of k is given by

$$k = 1.3805 \times 10^{-23} \text{ J/K}$$

Substituting the value of

$$\beta = \frac{1}{kT}$$

in all the equations, we get

$$Z = \sum_i g_i\, e^{-(Ei/kT)} \quad ...(v)$$

$$n_i = \frac{N}{Z} \sum_i g_i\, e^{-(Ei/kT)} \quad ...(vi)$$

This equation represents the Maxwell-Boltzmann distribution law in terms of the temperature of the system.

As $\quad \beta = \frac{1}{kT}$

$$d\beta = -\frac{dT}{kT^2}$$

Substituting this value in equation (ii)

$$\therefore \quad U = (kNT^2)\frac{d}{dT}(\log_e Z) \quad ...(vii)$$

Also, from equation (iii)

$$E_{av} = (kT^2)\frac{d}{dT}(\log_e Z) \quad ...(viii)$$

Equation (viii) gives the relation between the average energy of the particle and its temperature under equilibrium position. Hence, the temperature of a system in statistical equilibrium is the physical quantity related to the average energy of the particle of the system.

FERMI-DIRAC DISTRIBUTION LAW

Consider a system consisting of a large number of particles. It is assumed:

(i) The particles are *identical* and *indistinguishable*, and

(ii) The particles obey exclusion principle. It means that no two particles can have the same dynamical state and the wave function of the whole system must be antisymmetric. The particle satisfying these conditions are called *fermions*. In general, all fundamental particles with spin are fermions.

In quantum statistics the intrinsic probability g_i is governed by the different quantum states relating to a given energy *i.e.*, the degeneracy of the energy state. Each quantum state corresponds to a particle wave function. The Wave functions are determined by each of the possible arrangements of quantum number corresponding to a given energy level. For particles with spin 1/2, in the absence of magnetic forces, each of the particle may be in the energy states with spin +1/2 or –1/2. Hence the intrinsic probability g_i in this case is 2. For motion in a central field, the energy of the particle is independent of the orientation of the orbital angular momentum Due to this, a degeneracy of $(2l + 1)$ is introduced and it is the value of g for that particular energy state. If the particles possess spin, the total degeneracy = $2(2l + 1)$. As no two particles can be in the same energy state having the same quantum number, the instrinsic probabilities g_i's give the maximum number of particles that can be accommodated in a particular energy level without violating the exclusion principle *i.e.*,

$$n_i \leq g_i$$

It means that the n_i value for a given distribution does not exceed the corresponding value of g_i.

Let n_i, be the number of particles for energy level E_i The first particle can be placed in any one of the available g_i states *i.e.*, this particle can be assigned to any of the gi sets of quantum numbers. Thus, the first particle can be distributed in g_i different ways. Similarly, the second particle can be arranged in $(g_i - 1)$ different ways and the process continues.

Thus, the total number of different ways of arranging n_i particles among the available g_i states with energy level E_i is

$$= g_i(g_i - 1)(g_i - 2) ......[g_i - (n_i - 1)]$$

$$= \frac{g_i!}{(g_i - n_i)!} \quad ...(i)$$

Further, if the particles are taken to be indistinguishable, it will not be possible to detect any difference when n_i particles are reshuffled into different states occupied by them in the energy level E_i. Therefore, the total number of different and distinguishable ways is,

$$= \frac{g_i!}{n_i!(g_i - n_i)!} \quad ...(ii)$$

Therefore, the total number of different and distinguishable ways of getting the distribution n_1, n_2, n_3, etc., among the various energy levels, E_1, E_2, E_3...,etc.. can be obtained by multiplying the various factors.

$$\therefore \quad P = \frac{g_1!}{n_1!(g_1 - n_1)!} \cdot \frac{g_{2i}!}{n_2!(g_2 - n_2)!} ....$$

$$P = \prod_i \frac{g_i!}{n_i!(g_i - n_i)!} \quad ...(iii)$$

The most probable distribution can by obtained by evaluating the maximum value of $\log_e$, P in equation (iii).

This should also satisfy the condition that

$$\sum_i n_i = N$$

and $\sum_i n_i E_i = U.$

According to Stirling's approximation

$$\log_e x! = x \log_e x - x$$

From equation (iii), applying Stirling's approximation,

$$\log_e P = \sum_i [(g_i \log_e g_i - g_1) - (n_i \log_e n_i - n_i) - [(g_i - n_1) \log_e (g_i - n_1) - (g_i - n_1)]$$

$$\log_e P = \sum_i [g_i \log_e g_i - n_1 - \log_e n_i - (g_i - n_i) \log_e (g_i - n_1)] \quad ....(iv)$$

Differentiating equation (iv)

$$-d (\log_e P) = \sum_i [\log_e n_i - \log_e (g_i - n_i)] dn_i$$

To obtain the maximum value of P,

$$d(\log_e P) = 0$$

$$\therefore \quad \sum_i [\log_e n_i - \log_e (g_i - n_i)] dn_i = 0 \qquad \text{...(v)}$$

But $$\sum_i dn_i = 0 \qquad \text{...(vi)}$$

and $$\sum_i E_i dn_i = 0 \qquad \text{...(vii)}$$

Multiplying (vi) by α and (vii) by β and adding to equation (v), we get

$$\sum_i [\log_e n_i - \log_e (g_i - n_1) + \alpha + \beta E_i] dn_i = 0$$

The equilibrium distribution is possible if

$$\therefore \quad \log_e n_i - \log_e (g_i - n_1) + \alpha + \beta E_i = 0 \qquad \text{...(viii)}$$

$$\log_e \left(\frac{n_i}{g_i - n_i} \right) = -\alpha + \beta E_i$$

$$\frac{n_i}{g_i - n_i} = e^{-\alpha - \beta Ei}$$

$$\frac{g_i - n_i}{n_i} = e^{\alpha + \beta Ei}$$

$$\frac{g_i}{n_i} - 1 = e^{\alpha + \beta Ei}$$

$$\frac{g_i}{n_i} = (e^{\alpha + \beta Ei}) + 1$$

$$n_i = \left(\frac{g_i}{e^{\alpha + \beta E_i}} \right) + 1 \qquad \text{...(ix)}$$

Equation (ix) represents the Fermi-Dirac distribution law. The parameter has the same role as in the case of Maxwell-Boltzmann distribution law *i.e.*, for the system consisting of fermions in statistical equilibrium.

$$\beta = \frac{1}{kT}$$

$$\therefore \quad n_i = \frac{g_i}{[e^{(\alpha + E_i/kT)}] + 1}$$

In most cases the value of a is negative and is taken to be equal to

$$\frac{-E_F}{kT}$$

$$n_i = \frac{g_i}{\left[e^{(E_i - E_F)/kT}\right] + 1}$$

The value of E_F is positive and is independent of temperature.

For T = 0, all the energy states are fully occupied and $n_i = g_i$ All the states with $E > E_F$, are empty *i.e.*, $n_i = 0$.

$$\left[e^{(E_i - E_F)/kT}\right] = \begin{cases} 0 \text{ for } E_i - E_F < 0 \\ \infty \text{ for } E_i - E_F > 0 \end{cases}$$

In the case of Fermi-Dirac statistics, the accumulation of particles at the ground level is not allowed and at temperature T = 0, the particles occupy the lowest energy levels upto E_F. Here, the energy E_F gives the indication of the maximum energy of the fermions in the system. By is also called Fermi energy. For higher temperatures, the particles, occupy higher energy states greater than E_F. The curves indicate that only those fermions with energies close to E_F can move into unoccupied higher energy states. If $k0_F = E_F$, the temperature θ_F is called the Fermi temperature.

MAXWELL-BOLTZMANN DISTRIBUTION AND IDEAL GAS

Most of the gases obey Maxwell-Boltzmann distribution over a wide range of temperature. Consider a gas having mono-atomic molecules. It is assumed that the gas is ideal and has only kinetic energy of translation. Intermolecular attraction is assumed to be absent. If the kinetic energy is not quantized but considered to possess continuous values of energy, then the partition function Z can be written as

$$Z = \int_0^\infty e^{-(E/kT)} g(E)dE$$

Here g_i is replaced by g (E) dE

$$\text{But} \quad g(E)dE = \left(\frac{4\pi V(2m^3)^{1/2}}{h^3}\right)E^{1/2}dE$$

Here V is the volume occupied by the gas and h is Planck's constant.

$$Z = \frac{4pV(2m^3)^{1/2}}{h^3}\int_0^\infty E^{1/2}e^{-(E/kT)}dE$$

$$Z = \frac{4\pi V(2m^3)^{1/2}}{h^3}\left[\frac{\sqrt{\pi(kT)^3}}{2}\right]$$

$$Z = \frac{V(2\pi mkT)^{3/2}}{h^3} \qquad ...(i)$$

Equation (i) represents the partition function for an ideal monoatomic gas in terms of the volume and temperature of the gas.

Taking logarithms on both sides

$$\log_e Z = \log_e\left[\frac{(2\pi m)^{3/2} V}{h^3}\right] + \log_e (kT)^{3/2}$$

$$\log_e Z = C + \frac{3}{2}\log_e kT \qquad ...(ii)$$

Also $\quad E_{av} = (kT^2)\frac{d}{dT}(\log_e Z)$

$$E_{av} = (kT^2)\frac{d}{dT}\left[C + \frac{3}{2}\log_e kT\right]$$

$$= (kT^2)\frac{d}{dT}\left[C + \frac{3}{2}\log_e k + \frac{3}{2}\log_e T\right]$$

$$E_{av} = \frac{3}{2}kT \qquad ...(iii)$$

The total energy,

$$U = NE_{av}$$

$$U = N\left(\frac{3}{2}kT\right)$$

$$U = \frac{3}{2}NkT. \qquad ...(iv)$$

Equation (iv) shows that the internal energy of an ideal monoatomic gas depends only on its temperature. The same relation does not however, hold good for real gases. In the case of real gases the internal energy is partly potential and partly kinetic. The total energy depends on the volume of the gas.

ELECTRON GAS

Electrons in a metal belong to a most characteristic system of fermions because electrons obey the exclusion principle. For electrons in a metal, the energy levels are grouped in bands. Practically at all temperatures, the lower level energy bands are filled with electrons. The upper level energy bands are only partially filled with electrons. The distribution of electrons is to be considered only in the upper bands called the conduction band.

The zero energy level is taken at the lowest level of the conduction band. It is also assumed that the electrons have free movement within the conductor, provided the energy associated with the electrons is of the order of upper level energy bands.

As the energy of the electron in the conduction band is continuous, the term g_i is replaced by g(E) dE. Here dn electrons have energy in the range E and E + dE. According to Fermi-Dirac distribution law

$$n_i = \frac{g_i}{\left[e^{(E_i - E_F)/kT)}\right] + 1} \quad \text{...(i)}$$

Substituting the value of g_i = g (E) dB in equation (i) and replacing n_i by dn

$$dn = \frac{g(E)dE}{\left[e^{(E_i - E_F)/kT)}\right] + 1}$$

As an electron has the spin ±1/2, the total number of states in the sphere is twice. $V/(2\pi)^3$ refers to the translational states per unit volume in the Fermi space Fermi sphere of radius kF has the total number of particles accommodated,

$$n = \frac{2\left[\frac{V}{(2\pi)^3}\right]\left(\frac{4}{3}\pi(kF)^3\right)}{\left[e^{(E_i - E_F)/kT)}\right] + 1}$$

Here $$E = \frac{p^2}{2m} = \frac{\left(\frac{h}{2\pi}\right)^2 (kF)^2}{2m}$$

$$\therefore \quad kF = \frac{p}{h/2\pi} = \frac{2\pi p}{h}$$

$$\therefore \quad n = \frac{2\left[\frac{V}{(2\pi)^3}\right]\left(\frac{4}{3}\pi\left(\frac{2\pi p}{h}\right)^3\right)}{\left[e^{(E_i - E_F)/kT)}\right] + 1}$$

But $p = (2mE)^{1/2}$

$$\therefore \quad n = \frac{2\left[\frac{V}{(2\pi)^3}\right]\left[\left(\frac{4}{3}\pi\left(\frac{2\pi p}{h}\right)^3(2mE)^{3/2}\right)\right]}{\left[e^{(E_i - E_F)/kT)}\right]+1}$$

$$n = \frac{\left(\frac{8\pi V}{(3h^3}\right)(2m)((2m)^{1/2}E^{3/2}}{\left[e^{(E_i - E_F)/kT)}\right]+1}$$

Differentiating

$$dn = \frac{\left(\frac{8\pi V}{(3h^3}\right)(2m)((2m)^{1/2}\times\frac{3}{2}E^{1/2}dE}{\left[e^{(E_i - E_F)/kT)}\right]+1}$$

$$\frac{dn}{dE} = \frac{\left[\frac{8\pi V(2m^3)^{1/2}}{(3h^3}\right]E^{1/2}}{\left[e^{(E_i - E_F)/kT)}\right]+1} \quad \text{...(ii)}$$

Equation (ii) represents the energy distribution for free electrons. This is also called Fermi-Dirac formula of free fermions.

At $\quad T = 0K,$

$$p = (2m\ E_F)^{1/2} \quad \text{...(iii)}$$

and $\quad N = \left(\frac{2V}{h^3}\right)\left(\frac{4}{3}\pi p^3\right)$

or $\quad P = \left(\frac{3Nh^3}{8\pi V}\right)^{1/3} \quad \text{...(iv)}$

Equating (iii) and (iv)

$$(2mE_F)^{1/2} = \left(\frac{3Nh^3}{8\pi V}\right)^{1/3}$$

Squaring

$$2mE_F)^{1/2} = \left(\frac{3Nh^3}{8\pi V}\right)^{2/3}$$

or $$E_F = \frac{h^2}{8m}\left(\frac{3N}{\pi V}\right)^{2/3} \qquad ...(v)$$

Knowing the value of N/V *i.e.*, the number of free electrons per unit volume, the Fermi energy for electrons in a metal can be obtained.

For silver, the value $\frac{N}{V}$

$$= 5.86 \times 10^{28} \text{ electrons/m}^3$$

$$\therefore \quad E_F = \left(\frac{(6.624\times10^{-34})^2}{8\times9\times10^{-31}}\right)\left(\frac{3\times5.86\times10^{28}}{3.14}\right)^{2/3}$$

$$E_F = 9 \times 10^{-19}\text{J}.$$

But $$1\text{eV} = 1.6 \times 10^{-19}\text{J}$$

$$\therefore \quad E_F = \frac{9\times10^{-19}}{1.6\times10^{-19}}$$

$$E_F = 5.625 \text{ eV}.$$

It means that the maximum kinetic energy of free electrons in silver at absolute zero temperature is 5.6 eV.

The Fermi temperature θ_F is given by

$$\theta_F = \frac{E_F}{k}$$

As E_F practically independent of temperature, the value θ_F is fixed for a given metal. The values of θ_F and E_F are given in the following table.

Metals	*E_F (electron volts)*	*θ_F(K)*
Potassium	2.14	2.4×10^4
Sodium	312	3.7×10^4
Lithium	4.72	6.5×10^4
Silver	5.51	6.4×10^4
Gold	5.64	6.4×10^4
Copper	7.04	8.2×10^4

DISTRIBUTION LAW OF BOSE-EINSTEIN

Bose-Einstein distribution is applied to systems composed of identical and indistinguishable particles that are not restricted by the exclusion principle. In such systems, there is no limit to the number of particles occupying a particular quantum state. These particles are called *bosons*. Their spin is zero or 1. Mesons and helium nuclei are examples of bosons.

The Bose-Einstein statistics values of g_i refer to degeneracy of each energy level. Suppose, that n_i particles are arranged in a row and distributed among g_i quantum states with $(g_i - 1)$ partitions in between.

The total number of possible arrangements of particles and partitions is equal to the total number of permutations of $(n_i + g_i - 1)$ objects in a row. Therefore the total possible ways of arranging n_i particles with $g_i - 1$ partitions

$$= (n_i + g_i - 1)!$$

As the particles are identical and indistinguishable the possible number of distinct arrangements

$$= \frac{(n_i + g_i - 1)!}{n_i!(g_i - 1)!}.$$

The total number of distinguishable and distinct ways of arranging N particles in all the available energy states is given by

$$P = \frac{(n_1 + g_1 - 1)!}{n_1!(g_1 - 1)!} \times \frac{(n_2 + g_2 - 1)!}{n_2!(g_2 - 1)!} \times \ldots$$

$$P = \prod_i \frac{(n_i + g_i - 1)!}{n_i!(g_i - 1)!} \qquad \text{...(i)}$$

The most probable distribution can be obtained by finding the maximum value of $\log_e$, P.

According to Stirling's approximation

$$\log_e x! = x \log_e x - x$$

Also $$\sum_i n_i = N \qquad \text{...(ii)}$$

and $$\sum_i n_i E_i = U \qquad \text{...(iii)}$$

From equation (i), applying Stirling's approximation

$$\log_e P = \sum_i [\log_e (n_i + g_i - 1) - \log_e n_i ! \log_e (g_i - 1)!]$$

$$\log_e P = \sum_i [(n_i + g_i - 1) \log_e (n_i + g_i - 1) - (n_i + g_i - 1)$$

$$- (n_i \log_e n_i - n_i) - (g_i - 1) \log_e (g_i - 1) - (g_i - 1)]$$

$$\log_e P = \sum_i [(n_i + g_i - 1) \log_e(n_i + g_i - 1)$$
$$- n_i \log_e - n_i - (g_i - 1) \log_e(g_i - 1)] \quad ...(iv)$$

The maximum value of P is obtained by taking

$$d(\log_e P) = 0.$$

Differentiating equation P.

$$d(\log_e P) = \sum_i [\log_e (n_i + g_i - 1)\, dn_i - \log_e n_i - dn_i] = 0$$

$$\therefore \quad -d(\log_e P) = \sum_i [-\log_e (n_i + g_i - 1)\, dn_i + \log_e n_i dn_i] = 0$$

or

$$\sum_i [-\log_e (n_i + g_i - 1) + \log_e n_i] dn_i = 0 \quad ...(v)$$

As the total number of particles and total energy are constants, we have,

$$\sum_i dn_i = 0 \quad ...(vi)$$

$$\sum_i E_i dn_i = 0 \quad ...(vii)$$

Multiplying (vi) by α and equation (vii) by β and adding to equation (v), we get

$$\sum_i [-\log_e (n_i + g_i - 1) + \log_e n_i + \alpha + \beta E_i] dn_i = 0$$

$$\therefore \quad -\log_e (n_i + g_i - 1) + \log_e n_i + \alpha + \beta E_i = 0$$

Taking $n_i + g_i$ very large as compared to 1, the quantity 1 can be neglected.

$$\therefore \quad -\log_e (n_i + g_i) + \log_e n_i + \alpha + \beta E_i = 0$$

$$\log_e \left(\frac{n_i}{ng_i + g_i} \right) = -\alpha - \beta E_i$$

$$\frac{n_i}{n_i + g_i} = e^{-\alpha - \beta Ei}$$

$$\frac{n_i + g_i}{n_i} = e^{\alpha + \beta Ei}$$

$$1 + \frac{g_i}{n_i} = e^{\alpha + \beta Ei}$$

$$\frac{g_i}{n_i} = (e^{\alpha + \beta Ei}) - 1$$

$$n_i = \left(\frac{g_i}{e^{\alpha+\beta E_i}}\right) - 1 \qquad \text{...(viii)}$$

Equation (ix) represents the Bose-Einstein distribution law.

Taking $\beta = \dfrac{1}{kT}$

$$\therefore \quad n_i = \frac{g_i}{[e^{(\alpha+E_i/kT)}]-1}$$

The value of the constant a is governed by the equation $\Sigma n_i = N$. As n_i cannot be negative, α must always have a positive value. The distribution of the particles for different energy levels

SOME DEFINITIONS

It would be appropriate to discuss, at some length, the physical entities involved, in understanding the statistical problems.

A Physical Quantity

Essentially, we have two kinds of physical quantities (a) the intensive quantities that do not depend on the mass of the system, *e.g.*, P and T, and (b) the extensive quantities which depend on the mass of system, *e.g.*, S, V, E, etc.

System

It is a finite region in the space where the experimental body is kept. It Is obligatory for the system to have a finite mass. The system can be a closed one (may have a real physical boundary) or may have an imaginary boundary in space. Hence it can be penetrable or Impenetrable to the transport of energy. The properties of the particles in the system define the properties of the system.

Microstate

Each macrostate can have several microstates associated with it. Microstate is that state of the system where the individuality of the particles is not ignored. Hence the microstates are affected by change in position, momentum, energy (and mutual interaction) of individual particles. These states, however, cannot be identified by any physical experiment. Following table will help in understanding the difference between macro and microstates. We take two boxes A and B and four distinguishable particles a, b, c, d, and investigate their distribution.

It is evident from part B of the table that when the individuality of the particles is ignored, there remains no difference between the number of macrostates and associated microstates. By definition, given above, such a distribution will not give statistical probability or correct probable distribution (last column). Another illustration for the above table is to distribute 4 distinct books on 2 shelves. We have a total of 16 distributions as in part A of the table but if we distribute 4 identical books on same 2 shelves we have only 5 arrangements.

From the table, we see that for the most probable distribution (2, 2) we have 6 microstates. In other words, the macro distribution is most probable when number of microstates involved is maximum. In actual physical problems, we deal with extremely large number of particles as $\approx 10m^{-3}$ whence the number of microstates can be unimaginably large.

Macrostate

It is that state of the system which can be perceived by experiments. It is an isolated state of the system (not influenced by external agencies) and can be defined by certain value of pressure P, temperature T, and volume V. A macrostate can also be defined in terms of its microstructure, *e.g.*, by the specification of number of molecules (or phase-points in each cell of the phase space). Let us consider the distribution of 8 particles in two boxes. We can have (0, 8), (1, 7), (2, 6)... (7, 1) and (8, 0) distributions. These 9 distributions form 9 macrostates. Here we may note that a macrostate does not distinguish individuality of particles. Hence, for n particles we have (n + 1) macrostates when the distribution of particles is restricted in 2 boxes. This number will increase if the number of boxes is increased.

Phase Space

In order to specify the state of the gas from the molecular point of view, we shall know the position and momentum of each of its molecules, *i.e.*, we must specify six quantities x, y, z. p_x, p_y, p_z for each of the molecules. In three-dimensional space, we can easily consider an element of volume dx dy dz. Similarly, we can plot momentum components p_x, p_y and p_z in three mutually perpendicular directions and again consider three-dimensional volume. In other words, let us consider a particle with only a motion of translation. Then, classically we need three positional and three momentum coordinates to describe its state, (see point P in Fig. 2.3). Just as the space is represented by a three dimensional positional coordinate system, we can also represent momentum space by a three-dimensional momentum coordinate system. The state of the system then can be defined by six-dimensional positional-space and momentum space (phase) system. The two spaces together, defined by 6-dimensional (3 space and 3 momentum) hypothetical coordinate systems, is called phase space. Phase-space is purely a mathematical concept and cannot be depicted by a diagram like. So, phase-space is just a hypothetical space occupied by an infinite set of such six coordinate points. With the passage of time the point P (known as phase-point) traces out a certain trajectory known as phase-path.

As the position and momentum, in translated motion, change with lime, the phase-path may be quite complicated. Such 6-dimensional (6-d) phase space is also known as the μ-space. A large number of 6-d cells, each with a volume of $d\tau = (dx\ dy\ dz)\ (dp_x\ dp_y\ dp_z) = dV.\ dp$, together form the μ-space.

Let us now pass on to a many particle system. Let the system has N particles. Then, to define their state we need 3N positional and 3N momentum coordinates in a 6-d phase-space. We can consider this problem in another way:

A single point in 3-d space represents 3 coordinates; in 6-d space represents 6 coordinates; in n-d space represents n coordinates. Hence in 6N-d space a single point corresponds to 6N coordinates. If N be the number of molecules in a kilogram-mole (or a Mole), the 6N-dimensional phase-space is called r-space (or y-space). We can generalise this representation by assuming that the system has N particles each having f degrees of freedom the Γ- space will then be 2Nf dimensional, having Nf positional coordinates (q_i, i = 1 to Nf) and Nf momentum coordinates (P_i, i = 1 to Nf).

Phase-cells

The concept of phase space enables us to understand the state of micro particles -A their position and energy. When a particle changes its momentum and position with time, there will be infinite points in the phase-space which will express the state of this moving particle. Again in a finite volume of phase space, we have infinite microstates available for the particle. Hence, we can divide the whole of the phase space into finite sized small cells. The volume of the cell, however, cannot indefinitely be small.

Accessible Microstates

Those microstates which follow stipulated conditions (like number of particles is conserved, energy is conserved, system as finite extension) are called the accessible microstates.

The State of Equilibrium

It is the steady state of an isolated system where the values of P, V and T do not change with the passage of time. In such systems, no energy is given to or taken out of it. In statistical terms, equilibrium state corresponds to that macrostate which has maximum microstates. One must, however, remember that an isolated system, in equilibrium, is equally likely to be in any of its accessible microstate.

Events

When an operation is performed or is allowed to take place, the occurrence of certain complexion, configuration or arrangement of the system, at one such trial is called an event. When number of trials is increased the probability of certain event may increase.

STATISTICS APPLIED TO MOLECULES

A monatomic perfect gas. The state of each molecule of the gas in the μ-space is represented by a point (and by a cell in quantum mechanical terms). But, since the gas molecules are in constant motion, their state will be changing continuously in space and, due to collisions (and also mutual interactions), there is a change .in momentum state also. The trajectory of such a molecule constant energy surface given by

$$E = \frac{p_x^2 + p_y^2 + p_z^2}{2m} = V(q_x, q_y, q_z)$$

The real gas molecules, however, are endowed with mutual interactions in addition to usual collisions, hence their state is better represented by a point in y-space. The rate of change of its state, then, depends on the quantum of interactions. The path of representative point in y-space is given by the relation

$$E = \sum_i T_i + \sum_i V_i + \sum_{i,j}{}' U_{ij}$$

where T_i, V_i and U_{ij} refer to kinetic energy, potential energy and energy of interaction of ith particle. The prime on the summation sign suggests that i = j is not allowed (which would mean self-interaction).

With this background and remembering that the number of molecules is very large in a macrosystem, we now proceed to investigate the distribution details.

MACROSCOPIC SYSTEMS CHARACTERISTIC FEATURES

Let us now discuss some of the characteristic features of the macrosystems. In doing so we will assume that it has a large number of microscopic particles.

Fluctuation

Let the isolated system has N molecules of a perfect gas and that it was left to itself for quite some time so that it is in equilibrium. Further, also consider that the system has been partitioned into two identical (also equal) parts by an imaginary wall, each part A and B has n and n molecules respectively such that n + n = N.

Now, as we have seen, for large values of N, $n \cong n^1 = N/2$

But the molecules of the gas are in a state of constant random mction. So that some of the molecules are leaving portion .A and moving into B and vice-versa. Thus, there is a continuous change in the number of molecules in each portion of the box. Time variation of this change is known as fluctuation. The state of normal fluctuation of an isolated system. Usually, N/2 – n is only a small number. For large deviation, the number of microstates (and hence the probability) reduce drastically. The number of microstates for different values of N and N/2 – n. For the sake of simplicity the deviation N/2– n is represented as a fraction of total number of particles. Value of R ($\cong P_x/P_{max}$) is also shown to arrive at qualitative results;

In actual physical problems the number of molecules are $\sim 10^{24} m^{-3}$ whence even 0.01% deviation also becomes quite improbable. This is the reason for quick diffusion and most equitable distribution of air molecules in the available space.

EQUILIBRIUM STATE OF THE SYSTEM

The macrostate of a gas at a given instant is obtained from the distribution of molecules at that instant. But, if the macrostate of the multimolecular system does not change with time, the isolated system is said to be in equilibrium. Statistically, in the equilibrium state, the probability of the isolated system being in different accessible states does not vary with time. It may be noted that every system has an inherent tendency to approach equilibrium. If, anyhow, a situation (like point t_1 in Fig. 3.2) is created so that the system goes into a higher fluctuation state (lower probability), the motion of the molecules will set in such a way as to bring the system back into normal fluctuation state, with the passage of time. The time taken by the system to attain normal state from its disturbed (excited) state is called the relaxation *time of decay*. It is very easy to perturb the normal state of the system by external effects (by pressure/wind; diffusion of perfume is one such example). But the system will again tend to attain uniform distribution/most probable distribution. To check whether the macrostate is in equilibrium we plot fluctuation versus time curve for a certain period τ. The experiment is repeated again and again. If the curves agree macrostate will be in equilibrium state. In other words, the time average of number of molecules in certain region of space at equilibrium

$$\overline{n(t)} = \frac{1}{\tau}\int_t^{t+\tau} n(t)dt.$$

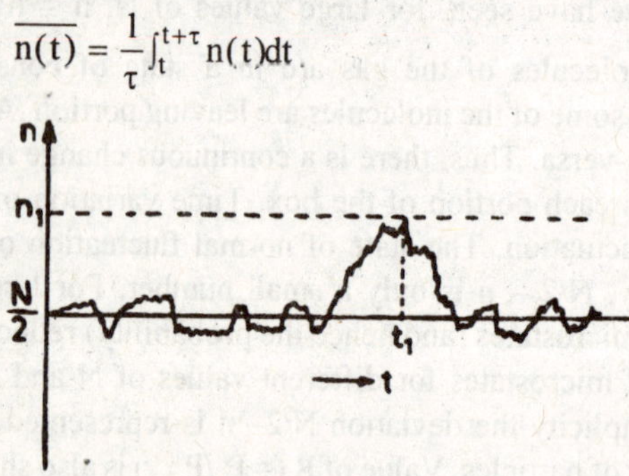

Fig. 3.2

GIBBS ENSEMBLE

The propeties of ensemble was put forth by Gibbs in 1902. In his view, it was a mathematical device which provided us with an opportunity to follow a probabilistic approach in defining the state of the system at a given time. Consider a system A such that a number of similar observations can be taken on it. Further, we assume that the system undergoes exactly similar changes in its state during each observation, *i.e.*, after every set of operations, the system returns back to its initial state (*i.e.*, the system is perfectly reversible). If this be the case, we can define the state of the system in terms of its statistical probability. But this is not practically feasible as:

(i) The number of molecules/particles in the system are so large that it is not possible to have the macrostate exactly the same as it was during the preceding observation,

(ii) Each macrostate has several microstates. We cannot ascertain their identity,

(iii) Each observation may take quite a long time whence for applying statistics, several observations may need several thousand years,

(iv) To manage and process a large number of data, over years together is not possible,

(v) The system may not show reversibility to different operations.

Ensemble was a satisfactory solution to it. An ensemble is a group of N_T identical systems. These N_T systems have the same configuration, energy and accessible states.

In other words, the ensemble has N_T isolated systems each of which are exact copies of the experimental system A Thus, we have a large number of systems that behave exactly in the same manner during a set of operations as the system A Then, if N_r systems out of N_T systems produce a particular result r, we have

$$P_r = N_r/N_T$$

as the probability for that result.

If we want to measure a particular property, say x, of the system

$$\langle x \rangle = \frac{n_1x_1 + n_2x_2 + ... + n_rx_r +}{n_1 + n_2 + ... + n_r +}$$

or $$\langle x \rangle = \frac{n_1x_1}{N_T} + \frac{n_2x_2}{N_T} + ... \qquad \because \Sigma n_i = N_T$$

or $$= P_1x_1 + P_2x_2 + ...P_rx_r + ...$$

i.e., $$= \Sigma P_rx_r$$

where x_r is the property of rth group of N_r similar systems and P_r is the probability for that result.

In this unique way Gibbs replaced the method of time averages (observations made on the same system at different times to obtain average) by ensemble average as shown above. The main advantage of the ensemble method is "the observations on all systems can be taken simultaneously under similar conditions", *e.g.*, test of quality of soaps of different brands will be quicker by millions of users.

SOME ENSEMBLES

It would be appropriate to classify the ensembles according of their dependency on macroparameters. Since an ensemble is an assembly of macroscopically identical but mutually independent systems, a change in macroparameter (like volume, pressure, energy, number of molecules, etc.,) may lead to another set of systems:

Canonical Ensemble

In this ensemble all its members are identical and remain in thermal equilibrium with the source (*i.e.*, members of the ensemble can exchange thermal energy with the reservoir). Thus all the members of a canonical ensemble have the same number of particles N, volume V and are at the same temperature T. A canonical ensemble. Its members are shown separated by impermeable but conducting rigid, walls and they are in thermal contact with the source. Gibbs gave a relation,

$$\rho(\varepsilon) = Ae^{-\varepsilon\tau}$$

for density of its phase points. A and T in the relation are constants. The ensemble can be viewed physically as a freezer containing water filled metal-ice cube tray.

Grand Canonical Ensemble

The members of this ensemble can interact thermally, mechanically and also diffusively with each other and with the reservoir. Hence they are at the same temperature of same volume and at same chemical potential such an ensemble. Each member here is separated by flexible, permeable and conducting walls, which help exchange mass and energy between its members and with the surroundings. The constituent molecules in the member systems are thus of different kinds (different chemical species). The density of phase points in such ensemble is

$$\rho(N) = e^{(\Omega+N\mu-\varepsilon)/\tau}$$

where, Ω, μ and τ are constants. A grain of sugar in boiling water can best be viewed as a grand canonical ensemble.

Uniform Ensemble

Those ensembles which have same density in phase space, are called *uniform ensembles*.

Hence, if ΔN represents the number of system, in volume element $d\Gamma$ we have, for a uniform ensemble,

$$\underset{d\Gamma \to 0}{\text{limit}} \frac{1}{d\Gamma} \frac{\Delta N}{N} = \text{constant}$$

Microcanonical Ensemble

In these ensembles the energy of their member systems remain constant, *i.e.*, they do not interact with each other or with the surroundings. Thus, they are free from external effects. The ensemble is thus conceptually an idealised one and can be represented by a function that depends only on the energy. An ensemble, here ε and V are constant. Every member is separated from the other by rigid, impermeable and insulated walls. The density of phase points within each member system is uniform and is zero at the walls. Physically, we may view the microcanonical ensemble as a perfect insulating enclosure encompassing several isolated systems. It is suitable for ideal gases.

SOLAR CONSTANT

The source of heat (sun) radiations and it emits heat radiations in all directions. The earth receives only a fraction of the energy emitted by the sun. The atmosphere also absorbs a part of the heat radiations and air, clouds, dust particles etc., in the atmosphere scatter the heat and light radiations falling on them. From the quantity of heat radiations received by the earth, it is possible to estimate the temperature of the sun. Therefore, to determine the value of a constant, called solar constant, certain ideal conditions are taken into consideration. *It is the amount of heat energy (radiation) absorbed per minute by one sq cm of a perfectly black body surface placed at a mean distance of the earth from the sun, in the absence of the atmosphere, the surface being held perpendicular to the sun's rays.*

The instruments used to measure the solar constant are called *pyrheliometers*. The heat energy absorbed by a known area in a fixed time is found with the help of the pyrheliometers. To eliminate the effects of absorption by the atmosphere, the value of the solar constant is found at various altitudes of the sun on the same day under similar sky conditions. If S is the observed solar constant, S_0 the true solar constant and Z the altitude (angular elevation) of the sun, then

$$S = S_0 \, a^{\sec} z \quad \text{...(i)}$$

$$\log S = \log S_0 + \sec Z \log a \quad \text{...(ii)}$$

Here a is a constant.

A graph is plotted between log S along the y-axis and sec Z along the x-axis. The graph is a straight line. Produce the graph to meet the y-axis. The intercept on the y-axis gives log S_0.

From the value of log S_0, the value of S_0 the solar constant can be calculated. The value obtained varies between 1.90 and 2.60 calories per sq cm per minute.

Temperature of the Sun

The sun consists of a central hot pardon surrounded by the photosphere. The central portion has a temperature of the order of 10^7 K. The photosphere has a temperature of about 6000 K. This temperature is also called the effective temperature of the sun. Considering the sun as a perfect black body radiator, the temperature of the sun can be calculated.

Let the mean distance of the sun from the earth be R and S the solar constant. Then, the total amount' of heat energy received by the sphere of radius R in one minute = $4\pi R^2S$.

If r is the radius of the sun, then the amount of heat energy radiated by 1 sq cm surface of the sun in one minute

$$E = \frac{4\pi R^2 S}{4\pi r^2} = \left(\frac{R}{r}\right)^2 \times S$$

Taking $R = 148.48 \times 10^7$ km

$r = 6.928 \times 10^6$ km

The mean value of

$S = 1.94$ cals per cm^2 per minute.

$$E = \left(\frac{148.48\times10^7}{6.928\times10^5}\right) \times \frac{1.94}{60} \text{ cal per second} \quad ...(i)$$

Also $E = \sigma T^4$

But $\sigma = 5.75 \times 10^{-5}$ ergs per cm^2 per second

$$= \frac{5.75\times10^{-5}}{4.2\times10^7} \text{ cal per cm}^2 \text{ per second}$$

$$E = \frac{5.75\times10^{-5}}{4.2\times10^7}.T^4 \quad ...(ii)$$

Equating (i) and (ii)

$$\left(\frac{5.75\times10^{-5}}{4.2\times10^{7}}\right)T^4 = \left(\frac{148.48\times10^{7}}{6.928\times10^{5}}\right)\times\frac{1.94}{60}$$

$$T = 5730 \text{ K.} \qquad \text{...(iii)}$$

This temperature gives the effective temperature of the sun acting as a black body radiator. The actual temperature of the sun is higher than this value. The temperature of the sun is usually taken as 6000 K.

Temperature of the sun can also be calculated from Wien's displacement law,

$$\lambda_{max}T = 0.2892$$

The wavelength of the radiations for which the energy is maximum in the spectrum is 4900×10^{-8} cm.

Substituting the value of λ_m, the value of T comes out to be 5902 K. This value is in agreement with the accepted value. Hence, the effective temperature of the sun (photosphere) is about 6000 K.

STEFAN'S LAW (DERIVATION OF MATHEMATICAL)

The fact that blade body radiations exert pressure similar to a gas, helps in applying thermodynamics to heat radiations.

Let ψ be the energy density of radiations inside a uniform temperature enclosure at temperature T. P is the pressure and V is the volume.

Applying the first law of thermodynamics

$$\delta H = dU + P.dV \qquad \text{...(i)}$$

Applying thermodynamical relation

$$\left(\frac{\partial H}{\partial V}\right)_r = \left(\frac{\partial P}{\partial T}\right)_v \qquad \text{...(ii)}$$

$$\left(\frac{\partial U + P\partial V}{\partial V}\right)_r = T\left(\frac{\partial P}{\partial T}\right)_v$$

$$\left(\frac{\partial U}{\partial V}\right)_r = T\left(\frac{\partial P}{\partial T}\right)_v = P \qquad \text{...(iii)}$$

Now $\qquad U = V\psi$

and $\qquad P = \frac{\psi}{3}$

or $$\left[\frac{\partial U}{\partial V}\right]_r = \psi$$

Here ψ is a function of temperature alone.

Substituting these values in equation (iii),

$$\psi = \frac{T}{3}\frac{d\psi}{dT} - \frac{\psi}{3}$$

$$\frac{4\psi}{3} = \frac{T\,d\psi}{3\,dT}$$

or $$\frac{d\psi}{\psi} = 4\frac{dT}{T}$$

Integrating, log ψ = 4 log T + constant

or $$\psi = KT^4 \qquad ...(iv)$$

Here K is a constant

Also the total rate of emission per unit area of a black body is proportional to the energy density.

$$\therefore \qquad R \propto \psi \propto T^4$$

$$\therefore \qquad R = \sigma T^4 \qquad ...(v)$$

where σ is Stefan's constant.

The value of Stefan's constant in C.G.S. system is 5.672×10^{-5} C.G.S. units and in M.K.S. system it is 5.672×10^{-8} M.K.S. units.

COOLING FROM STEFAN'S LAW (DERIVATION OF NEWTON'S LAW)

Stefan's law is applicable for all temperatures of a hot body. But Newton's Law is applicable when the difference of temperature between the hot body and the surrounding is small. Consider a hot body at a temperature T_1 placed in a uniform temperature enclosure at T_1. According to Stefan's law,

$$R = e\sigma(T_1^4 - T_1^4) \qquad ...(i)$$

Here e is the emissivity of the surface of the hot body

$$R = e\sigma(T_1 - T_2)(T_1^3 + T_1^2 T_2 + T_1 T_2^2 + T_2^3)$$

As $(T_1 - T_2)$ is small, T_1 can be taken approximately equal to T_2.

Then, $$R = e\sigma(T_1 - T_2)(T_2^3 + T_2^3 + T_2^3 + T_2^3)$$

$$R = 4e\sigma T_2^3((T_1 - T_2)$$

Taking $4e\sigma T_2^3 = k$

$$R = k(T_1 - T_2)$$

or $R \propto (T_1 - T_2)$. ...(ii)

This equation represents Newton's law of cooling and is true when the difference of temperature is small.

STEFAN'S LAW VERIFICATION

In 1897, Lummer and Pringsheim experimentally verified Stefan's law over a wide range of temperature (100°C to 1,300°C). The apparatus consisted of a black body C. For temperatures between 200°C and 600°C, a hollow copper sphere coated inside with platinum black was used. The fused nitrates of sodium and potassium having a melting point of 219°C were used as the bath surrounding the black body. For temperature between 900°C and 1,300°C, an iron cylinder coated inside with platinum black was used as a black body and it was enclosed in a double walled gas furnace. A thermocouple T was used as a thermometer. A bolometer B was used to measure the intensity of the emitted heat radiations. S_1, S_2 and S_3 were the water-cooling shutters. Another black body A at 100°C was used to standardize the bolometer.

The double walled vessel of the black body A contained boiling water at 100°C. The bolometer B was allowed to face the opening of the black body A and the shutter S_3 raised. The deflections in the galvanometer of the bolometer at various distances were noted and it was found that the deflection was inversely proportional to the square of the distance between the bolometer and the opening of the black body A. Thus, the deflection in the galvanometer was proportional to the intensity of heat radiations.

The shutter S_3 was closed and the bolometer B was allowed to face the opening of the black body C. The shutters of S_1 and S_2 were raised. The bath surrounding the black body was maintained at a constant temperature and the maximum deflection produced in the galvanometer of the bolometer was noted. Thus at various constant temperatures of the black body, corresponding to constant deflections (in the galvanometer of the bolometer) were observed. Then the data was reduced to a common arbitrary unit in terms of the total radiations from the black body A at 100°C.

Let θ be the deflection in the galvanometer, T_1 the temperature of the black body and T_2, the temperature at the entrance of the bolometer. It was found that

$$\theta \propto (T_1^4 - T_2^4)$$

But $\quad \theta \propto R$

$$\therefore \quad \theta \propto (T_1^4 - T_2^4)$$

This verifies Stefan's law.

Recently Coblentz has verified Stefan's law more accurately. He took an electrically heated black body whose temperature was measured by an accurate thermocouple. An absolute bolometer was used to measure the amount of heat radiations emitted by the black body. He was able to show the correctness of Stefan's law experimentally up to 1,600°C.

PHOTON GAS

The interaction of electromagnetic radiations with matter, led to the idea that electromagnetic radiations are composed of discrete energy particles called photons. Each photon has an energy $h\nu$ and momentum h/λ. Here ν is the frequency and λ is the wavelength of the radiations. The electromagnetic radiations trapped in a cavity and in thermal equilibrium with the walls of the cavity are termed as black body radiations. In the equilibrium condition, the black body radiations can be considered as the *photon gas.* It is assumed that the photons do not interact among themselves. The photons interact only with-the atoms of the walls of the cavity. It is further assumed that the photons are indistinguishable and many photons can have the same energy. Photons are taken as bosons and they obey Bose-Einstein statistics.

As the photons can either be emitted or absorbed by the atoms of the walls of the cavity, the number of photons is not constant *i.e.*, the condition $\sum_i dn_i = 0$ is no longer valid.

Due to this reason the value of α is equal to zero in the Bose-Einstein distribution law.

$$n_i = \frac{g_i}{[e^{(\alpha + E_i/kT)}]-1} \qquad ...(i)$$

Here $\quad \alpha = 0$

$$\therefore \quad n_i = \frac{g_i}{[e^{E_i/kT}]-1} \qquad ...(ii)$$

In case, the cavity is large as compared to the wavelength of the radiations, the energy spectrum of the photons is taken to be continuous. In this case, the energy difference between successive allowed energy value is very small. Thus, replacing g_i by g(E) dE, and n_i by dn

$$dn = \frac{g(E)dE}{\left(e^{E/kT}\right) - 1} \qquad \text{...(iii)}$$

As the energy of a photon, E = hν, the value of g(E) dE can be taken equal to g (ν) dν. The factor g(ν) dν corresponds to the number of oscillatory modes in the frequency range dν and relating to the energy range dE.

The number of states in a black body radiation in the frequency range ν and ν + dν can be obtained by calculating the spherical volume bound by the spheres of radii $\frac{h(\nu + d\nu)}{c}$ and $\frac{h\nu}{c}$

The volume of the spherical shell

$$= \frac{4}{3}\pi\frac{h^3}{c^3}(\nu + d\nu)^3 - \frac{4}{3}\pi\frac{h^3}{c^3}\nu^3$$

$$= \frac{4}{3}\pi\frac{h^3}{c^3}[\nu^3 + 3\nu^2 d\nu + ... - \nu^3]$$

$$= \frac{4}{3}\pi\frac{h^3}{c^3} \times 3\nu^2 d\nu$$

$$= 4\pi \frac{h^3}{c^3} \nu^2 d\nu \qquad \text{...(iv)}$$

The phase space has volume V = h^3 and there are two states of polarization for the radiation. The number of states in the black body radiation in the frequency range ν and ν + dν is given by

$$g(\nu)d\nu = \frac{8\pi\nu}{c^3}\nu^2 d\nu \qquad \text{...(v)}$$

$$dn = \left(\frac{8\pi V\nu^2 d\nu}{c^3}\right)\frac{1}{\left(e^{h\nu/kT}\right) - 1} \qquad \text{...(vi)}$$

For dn photons in the frequency range ν + dν and ν, the energy is equal to (hν) dn and the energy per unit volume

$$= \frac{(h\nu)dn}{\nu}$$

Therefore, the energy density distribution for black body radiation is given by

$$E(\nu) = \frac{(h\nu)dn}{V.d\nu} \qquad \text{...(vii)}$$

Substituting the value of dn from equation (vi), we get

$$E(\nu) = \frac{8\pi h\nu^2 d\nu.h\nu}{c^3 V.d\nu\left[\left(e^{h\nu/kT}\right)-1\right]}$$

$$E(\nu) = \left(\frac{8\pi h\nu^3}{c^3}\right)\left[\frac{1}{\left(e^{h\nu/kT}\right)-1}\right]^{d\nu} \qquad \text{...(viii)}$$

Equation (viii) represents Planck's radiation law for black body radiations.

From equation (viii)

$$E(\nu)d\nu = \frac{8\pi h\nu^3}{c^3}\left[\frac{1}{\left(e^{h\nu/kE}\right)-1}\right]^{d\nu}$$

But $\nu = \frac{c}{\lambda}$

or $d\nu = -\frac{c}{\lambda^2}d\lambda$

Neglecting the negative sign, we get

$$\therefore \quad E(\lambda)d\lambda = \frac{8\pi h\nu^3}{c^3} \times \frac{c}{\lambda^2}\left[\frac{1}{\left(e^{h\nu/kT}\right)-1}\right]^{d\lambda}$$

or

$$E(\lambda) = \frac{8\pi hc}{\lambda^5} \times \frac{c}{\lambda^2}\left[\frac{1}{\left(e^{h\nu/kT}\right)-1}\right] \qquad \text{...(ix)}$$

This gives the energy density for wavelength λ in the spectrum of the black body. Both the equations (viii) and (ix) represent Planck's radiation law for black body radiation.

GIBBS FUNCTION

The Gibbs function G of a system is given by,

$$G =. U - TS + PV$$

Consider a system that can do other forms of work, in addition to P.dV work, *e.g.*, a voltaric cell. In the case of a voltaic cell, the electrical work is –E.dI. Similarly for a magnetic material, the magnetic work is –m.dH. In general, the work will be given by P.dV plus a sum of terms each being the product of intensive variable (such as P, E or m) and the differential of an extensive variable (such as dV, dI and dH). In the case of a voltaic cell, the intensive variable is S and differential of extensive variable is dI.

Suppose, in general, in addition to PdV the intensive variable is y and differential of extensive variable is dx. The work done for any reversible process,

$$\delta W = PdV + ydx$$

$$W = \int_{V_1}^{V_2} PdV + \int_{x_1}^{x_2} ydx$$

Take $\int_{x_1}^{x_2} ydx = A$

Consider a process where the system works at constant pressure P_0 and the change in volume is $(V_2 - V_1)$

$$\therefore \quad \int_{V_1}^{V_2} PdV = P_0 [V_2 - V_1]$$

$$\therefore \quad W = P_0[V_2 - V_1] + A \qquad ...(i)$$

For a system that exchanges heat with a reservoir temperature

$$W < (U_1 - U_2) - T_0 (S_1 - S_2) \qquad ...(ii)$$

$$P_0[V_2 - V_1] + A < (U_1 - U_2) - T_0 (S_1 - S_2)$$

$$\text{or} \quad A < (U_1 - U_2) - T_0 (S_1 - S_2) + P_0[V_1 - V_2] \qquad ...(iii)$$

Consider a specific process, where the initial and the final states of the system and the surroundings are at the same temperature (T_0) and pressure (P_0)

$$To = T \text{ and } P_0 = P$$

From equation (iii)

$$A_{P,T} < (U_1 - U_2)_{P,T} - T(S_1 - S_2)_{P,T} + P[V_1 - V_2]_{P,T} \qquad ...(iv)$$

But, for the Gibbs function,

$$G = U - TS + PV \qquad ...(v)$$

Therefore, for two equilibrium states at the same pressure and temperature,

$$[G_1 - G_2)_{P,T} = (U_1 - U_2)_{P,T} - T[S_1 - S_2]_{P,T} + P[V_1 - V_2]_{P,T} \quad ...(vi)$$

From equations (iv) and (vi)

$$A_{P,T} < (G_1 - G_2)_{P,T} \quad (vii)$$

Thus, the difference between Gribbs function of a system between two equilibrium states sets the maximum limit to the work in addition to PdV work, provided the initial and the final states are at the same pressure and temperature and the system exchanges heat with a single heat reservoir. The work done will be maximum when the process is reversible. The process in this case will be isothermal–isobaric. If the process is irreversible, work done will be less than the maximum.

SOLVED EXAMPLES

Example 1:

Is the equation PV^γ = constant valid for all processes ? Explain your answer.

Solution:

The expression PV^γ = constant is valid only for adiabatic reversible volume change (*i.e.*, expansion or compression) involving an ideal gas. For adiabatic process, dq = 0.

Hence

$$dU = dw$$

$$C_V dT = - P_{ext}\ dV$$

For revisable process, $P_{ext} = P_{int}$. Now replacing P_{int} = nRT/V and separating the variables followed by integration, we get

$$\frac{Cv}{n}\int\frac{dT}{T} = - R\int\frac{dV}{V}$$

or $$C_{V.m} \ln\frac{T_2}{T_1} = - R \ln\frac{V_2}{V_1}$$

or $$\left(\frac{T_2}{T_1}\right)^{C_{v,m}} = \left(\frac{V_2}{V_1}\right)^{-R}$$

Replacing T_2/T_1 by $P_2\ V_2/P_1\ V_1$ and rearranging, the expression would yield

$$P_2V^{\gamma}_2 = P_1\ V^{\gamma}_1$$

that is $$PV^{\gamma} = \text{constant.}$$

Example 2:

Under what conditions does the adiabatic expansion of gas become a constant energy process ?

Solution:

If the adiabatic expansions done against a vacuum then in this process energy

On comparing the coefficients in Eqs (1) and (2), we have

$$\left(\frac{\partial E}{\partial T}\right)_V = T \quad ...(3)$$

and $$\left(\frac{\partial E}{\partial V}\right)_S = -P \quad ...(4)$$

Eq (4) can be written as

$$dE = T\ dS \quad ...(5)$$

For a given temperature, the above equation on integration becomes as follows :

$$\int dE = \int TdS + \text{Integration constant}$$

$$E = TS + A \quad ...(6)$$

or $$E = A - TS$$

where A is an integration constant. It is also known as Helmholtz free energy function.

(d) Enthalpy as a Function of S and P

As H is a function of S and P, we can write

$$H = f\ (S,\ P)$$

or $$dH = \left(\frac{\partial H}{\partial S}\right)_P dS + \left(\frac{\partial H}{\partial P}\right)_S dP \quad ...(1)$$

We also know, $dH = T\ dS + V\ dP$ ...(2)

On comparing the coefficients in Eqs (1) and (2), we get

$$\left(\frac{\partial H}{\partial S}\right)_P = T \quad ...(3)$$

$$\left(\frac{\partial H}{\partial P}\right)_S = V \qquad ...(4)$$

When Eq. (3) is integrated, we get a new function

$$\int dH = \int TdS + \text{constant}$$
$$= \int TdS + G$$

or $$H = TS + G$$

or $$G = H - TS$$

where G is an integration constant and is the same as the Gibbs free energy function.

Example 3:

Two moles of an ideal gas at 300 K expand reversibly and isothermally from 4×10^{2} m^3 to 8×10^{-2} m^3. Calculate the entropy change for the gas. How would you account for the fact that ΔS is not zero although the process is reversible ?

Solution:

If the process is reversible isothermal it means that $\Delta H = 0$, hence the entropy change is given as follows :

$$\Delta S_{sys} = \frac{Q_{rev}}{T} = \frac{W_{rev}}{T} = nR \ln\frac{V_2}{V_1}$$

$$= (2\text{ mol})\,(8.314\text{ JK}^{-1}\text{ mol}^{-1})\left(2.303 \log\frac{8 \times 10^{-2}}{4 \times 10^{-2}}\right)$$

$$= 2 \times 8.314 \times 2.303 \log 2\text{ JK}^{-1}$$

$$= 2 \times 8.314 \times 2.303 \times 0.3010\text{ JK}^{-1}$$

$$= 11.526579\text{ JK}^{-1}.$$

The total entropy change (system and its surroundings) will be zero for a reversible process.

Example 4:

Calculate the entropy changes involved in the conversion of 1 mole of ice at 0°C and 1 atm to liquid at 0°C and 1 atm [The enthalpy of fusion per mole of ice is 6008 J mol^{-1}].

Solution:

$$H_2O\ (S) \rightarrow H_2O\ (l)$$

$$\Delta S_{fus} = \frac{\Delta H_{fus}}{T_m}$$

$$= \frac{6008\ J\ mol^{-1}}{273\ K}$$

$$= 22.007326\ JK^{-1.}$$

Example 5:

Increase in volume of a gas from a given decrease of pressure is more in an isothermal expansion than in an adiabatic expansion. Explain.

Solution:

In an isothermal expansion, temperature remains the same whereas in adiabatic expansion temperature decreases. Since the final pressure is them same, it follows from Charles law ($V \propto T$, p constant) that

$$V_{iso} > V_{adi}$$

Since $\Delta V_{iso} = V_{iso} - V_{intial}$

$$\Delta V_{adi} = V_{adi} - V_{intial}$$

it follows that.

Example 6:

Two bodies at different temperatures are connected by a wire with infinitesimal thermal conductance so that heat flows infinitely slowly from the hot bodies to the cold body until they are at the same temperature. Is this process reversible? Explain.

Solution:

In a reversible process, at any stage, the external condition responsible for the process to occur differs from the internal condition by an infinitesimal amount. The given transfer of heat is not a reversible process as the two bodies are at different temperatures. Moreover, the reversible processes are characterized by the fact that when the system is restored to its original state by traversing the forward sequence of steps in the reverse order, then not only the system but also the surroundings are restored to their original states. In the given process, heat flow from a hot body to a cold body does not disturb the surroundings but if we wish to carry out the reverse process, the surroundings will have to do some work. Hence, the surroundings are not brought back to the original state in the reverse process. Thus, the given process is not a reversible one.

Example 7:

How would the energy of an ideal gas change if it is made to expand into vacuum at constant temperature ?

Solution:

For the expansion against vacuum $P_{opp} = 0$

Hence $w = -P_{opp} \Delta V = 0$

Since the expansion is at constant temperature no heat is supplied to the system, *i.e.*, $q = 0$. Now, from the first law of thermodynamics, we can write

$$\Delta E = q + w = 0$$

that is, the energy of an ideal gas does not change if it is made to expand into vacuum at constant temperature. This also follows from the fact that the energy of an ideal gas depends only on temperature and since temperature remains constant, its energy also remains constant.

Example 8:

Calculate the change of entropy for the process

$$H_2O\ (l,\ 373\ K) \rightarrow H_2O\ (v,\ 373\ K)$$

Here $\Delta H_{vap} = 40850\ J\ mol^{-1}$ *at 373 K.*

Solution:

$$\Delta S_{vap} = \frac{\Delta H_{vap}}{T_b}$$

$$= \frac{40850\ J\ mol^{-1}}{373\ K}$$

$$= 109.51742\ JK^{-1}\ mol^{-1}$$

But $\Delta S_{vap} = S\ (vapour) - S\ (liquid)$

$= 109.5\ JK^{-1}\ mol^{-1}$

$\therefore$ $S\ (vapour) = S\ (liquid) + 109.5\ JK^{-1}\ mol^{-1}$

Example 9:

Evaluate the entropy change for the following reversible process.

$$1 \text{ mole Sn } (\propto, 13^oC \rightleftharpoons 1 \text{ mole Sn } (b, 13^oC)$$

where $\Delta H_{trans} = 2090 \ J \ mol^{-1}$

Solution

$$T_{trans} = 273 + 13 = 286 \text{ K.}$$

$$\Delta S_{trans} = \frac{\Delta H_{trans}}{T_{trans}}$$

$$= \frac{2090 \text{ J mol}^{-1}}{286 \text{ K}}$$

$$= 7.3076923 \text{ JK}^{-1} \text{ mol}^{-1}$$

Example 10:

Calculate the entropy change when two moles of; an ideal gas expands reversibly from an initial volume of 2 dm³ to a total volume of 20 dm³ at a constant temperature of 298 K.

Solution:

We know, $\Delta S = nR \ln \left(\frac{V_2}{V_1}\right)$

$$= 2.303 \text{ nR} \log \left(\frac{V_2}{V_1}\right)$$

$$= 2.303 \times 2 \times 8.314 \text{ J JK}^{-1} \text{ mol}^{-1} \log \left(\frac{20\text{dm}^3}{2\text{dm}^3}\right)$$

$$= 2.303 \times 2 \times 8.314 \times \log \text{JK}^{-1} \text{ mol}^{-1}$$

$$= 38.294284 \text{ JK}^{-1} \text{ mol}^{-1}.$$

Example 11:

The mixing of gases is always accompanied by an increase in entropy. Show that in the formation of a binary mixture of two ideal gases the maximum entropy increase results when $X_1 = X_2 = 0.5$.

Solution:

For a binary ;mixture, the entropy per mole of the mixture formed is given by

$$\Delta S \text{ mixing} = -R[X_1 \ln X_1 + X_2 \ln X_2]$$
$$= -R[X_1 \ln X_1 + (1 - X_1) \ln (1 - X_1)]$$

For entropy of mixing to be maximum, the first derivative, $\frac{\delta(\Delta S \text{ mixing})}{\delta X_1}$ should be zero and the second derivative should be negative. Differentiating, ΔS mixing with respect to X_1 and equating it to zero yields

$$\frac{\delta(\Delta S \text{ mixing})}{\delta X_1} = -R\left[\ln X_1 + \frac{X_1}{X_1} + \frac{1-X_1}{1-X_1}(-1) + (-1)\ln(1-X_1)\right]$$
$$= 0$$
$$\ln X_1 + 1 - 1 - \ln(1 - X_1) = 0$$
$$\ln \frac{X_1}{1-X_1} = 0$$
$$X_1 = 1 - X_1$$
$$X_1 = 1/2.$$

Example 12:

Under what conditions does the isothermal expansion of a gas become a free expansion process?

Solution:

The isothermal expansion of a gas becomes a free expansion process only when it expands against a vacuum, *i.e.*, $p_{ext} = 0$. Moreover, the gas must be an ideal gas in which case ΔE = 0 at constant temperature.

Example 13:

Is every adiabatic system isolated? In what ways can the energy of an adiabatic system change?

Solution:

An isolated system is the one in which no interaction with its surroundings takes place. Neither energy nor matter can be transferred to or from it. An adiabatic system involves no exchange of heat from its surroundings, *i.e.* for such a system dq = 0 and, according to the first law of thermodynamics, dE =dw. Thus, the energy of an adiabatic system can be changed by either doing work on the system or by getting work done by the system.

Example 14:

One mole of a gas is expanded isothermally and reversibly at 27° C form a volume of 5 liters. Calculate q, w, ΔE and ΔH.

Solution:

We know

$$PV = nRT, \quad n = 1$$

State : Initial

$T_1 = 27 + 273.15 = 300.15$ K

$V_1 = 5$ litres

Final

$T_2 = T_1 = 300.15$K

$V_2 = 10$ litres

(1) Calculation of w :

$$w = \int_{V_1}^{V_2} P\,dV = \int_{V_1}^{V_2} \frac{RT}{V} dV = RT \ln \frac{V_2}{V_1}$$

$$= 2.303 \times 8.314 \times 300 \log (10/5)$$

$$= 2.303 \times 8.314 \times 300 \times \log^2$$

$$= 2.303 \times 8.314 \times 300 \times 0.3010 \quad [\log 2 = 0.3010]$$

$$w = 1729 \text{ J mol}^{-1}$$

(2) Calculation of ΔE :

$$\Delta E = E_2 - E_1$$

For isothermal-reversible expansion of an ideal gas $E_2 = E_1$ and therefore

$$\Delta E = 0.$$

(3) Calculation of q:

$$q = \Delta E + w \text{ (first law)}$$

$$= 0 + w$$

$$q = w = 1729 \text{ J}$$

(4) Calculation of ΔH

$$\Delta H = \Delta E + \Delta (PV) = \Delta E + \Delta(RT)$$

$$0 + 0 = 0$$

Since $\Delta E = 0$ and $\Delta(RT) = 0$; R and T being constant

Example 15:

Calculate the minimum work which must be done to compress 1/2 mole of oxygen at 300 k from a pressure of 2 atm to a pressure of 200 atm.

Solution:

As it is required to calculate the minimum work of compression, the change, is reversible, *i.e.*, the gas gets compressed by gradual increase of pressure so that the external pressure becomes only infinitesimally greater than the pressure of the gas, at each stage. Thus

$$w = \text{work done} = \int_{V_1}^{V_2} P\,dV$$

$$= 2.303 \text{ n RT} \log \frac{V_2}{V_1}$$

$$= 2.303 \text{ nRT} \log \frac{p_1}{P_2}$$

given date :

$$P_1 = 2 \text{ atm}$$

$$P_2 = 2 \text{ atm}$$

$$T = 300 \text{ K}$$

$$n = \frac{16}{32} = 0.5 \text{ mole.}$$

On substituting these values in the above equation, we get

$$w = 2.303 \times 0.5 \times 8.314 \times 300 \log \frac{2}{200}$$

$$= - 5744 \text{ J.}$$

Example 16:

Four moles of an ideal gas are held by a piston under 5 atm pressure and at 273.15 K. The pressure is suddenly released to 0.2 atm and the gas is allowed to expand isothermally. Calculate w, q, ΔE and ΔH for the process.

Solution:

As the pressure has been released suddenly, the expansion must be irreversible.

Here n = 4 moles; P_1 = 5 atm; $T_1 = T_2$ = 273.15 K.

(i) Work done irreversibly

$$w = nRT\left\{1 - \frac{P_2}{P_1}\right\}$$

$$= 4 \times 8.314 \times 273.15 \left\{1 - \frac{0.2}{5}\right\}$$

$$= 4 \times 8.314 \times 273.15 \times 0.96$$

$$= 8720 \text{ J.}$$

(ii) As the ideal gas expands isothermally

$$dT = 0 \text{ and hence } \Delta E = 0; \Delta H = 0.$$

(iii) First law gives

$$\Delta E = q - w = 0$$

$$\therefore q = w\ 8720 \text{ J.}$$

Example 17:

At low temperature, a nonideal gas follows the relation PV = RT – a/V where a = 0.3636 Nm⁴ mole⁻². Calculate the work done by one mole of this gas in expanding form 0.224 litre to 22.4 litre at 400 K. Compare this result with the work done in the corresponding expansion of an ideal gas.

Solution:

For an ideal gas, reversible work done

$$= 2.303 \text{ n RT} \log \frac{V_2}{V_1}$$

V_1 = 0.224 litres mole^{-1} = 22.4 × 10^{-5} m^3 mole^{-1}

P_2 = 22.4 litres mole^{-1} = 22.4 × 10^{-3} m^3 mole^{-1}

R = 8.314 J K^{-1} mole^{-1}

$$W_{rew} = (2.303)\ (1 \text{ mole})\ (8.314 \text{ J K}^{-1}.\text{mole}^{-1})\ (400 \text{ K}) \times \log \frac{22.4 \times 10^{-3}}{22.4 \times 10^{-5}}$$

$$= (2.303)\ (1 \text{ mole})\ (8.314 \text{ J K}^{-1} \text{mole}^{-1})\ (400 \text{ K}) \log 100$$

$= 15320 \text{ J mole}^{-1}$

$= 15.32 \text{ kJ mole}^{-1}$.

For the non-ideal gas PV = RT – a/V

$$\text{Therefore, the work done} = \int_{V_1}^{V_2} P\,dV = \int_{V_1}^{V_2} \left\{\frac{RT}{V} - \frac{a}{V_2}\right\} dV$$

$$= 2.303 \text{ RT} \log \frac{V_2}{V_1} + a\left\{\frac{1}{V_2} - \frac{1}{V_1}\right\}$$

The value of first part is from derived eq. (1) as $15.32 \text{ kJ mole}^{-1}$ and of another part can be derived as

$$0.3636 \text{ Nm}^4 \text{ mole}^{-2} \left\{\frac{1}{22.4 \times 10^{-3} \text{ m}^3 \text{ mole}^{-1}} - \frac{1}{22.4 \times 10^{-5} \text{ m}^3 \text{ mole}^{-1}}\right\}$$

$= 1.604 \text{ kJ mole}^{-1}$.

Thus work done $= 15.320 \text{ k Joule}^{-1} - 1.604 \text{ k Joule}^{-1}$

$= 13.716 \text{ kJ mole}^{-1}$.

Example 18:

What will be the work done when 65.38 g of zinc dissolves in hydrochloric acid in case of an open beaker and closed beaker at 300K.

Solution:

The process of dissolution of zinc in hydrochloric acid can be given as

$$Zn(s) + 2HCI\ (aq) = ZnCl_2(aq) + H_2(g)$$

From this equation it is clear that for each gram atom of zinc dissolved we will obtain 1 mole of Hydrogen gas. As a result of liberated gas, on the surrounding atmosphere the work performed will be equal to PΔV. If the initial volume of the system is neglected in comparison to the total volume of the gas produced and the gas shows ideal behaviour, then the work done in the open beaker is

$$W = P\Delta V \approx PV_{H2}$$

$= nH2\ RT \quad [\because T = 300 \text{ k } R = 8.314 \text{ Jk}^{-1} \text{ mole}^{-1}]$

$= (1 \text{ mole})\ (8.314 \text{ JK}^{-1} \text{ mol}^{-1})\ (300 \text{ K})$

$= 2.4942$ kJ.

If the reaction takes place in a closed beaker there is change in the volume *i.e.*, $\Delta V = 0$ hence work done is zero.

Example 19:

Calculate the work done when one mole of sulphur dioxide gas expands isothermally and reversibly at 300 k from 2.46 × 10 ³ m³ to 24.6 × 10⁻³ m³. Assuming that the gas obeys

(a) ideal gas equation and

(b) van der Waals equation.

$$\left\{\begin{array}{l} a = 0.6799 \text{ nm}^4 \text{ mole}^{-2} \\ b = 0.564 \times 10^{-3} \text{ mole}^{-1} \end{array}\right\}$$

Solution:

(a) The reversible work done by 1 mole of gas when it expands form

$$2.64 \times 10^{-3}\text{m}^3 \text{ to } 24.6 \times 10^{-3}\text{m}^3 = 2.303 \text{ RT log } \frac{V_2}{V_1}$$

$$= (2.303)\ (1 \text{ mole})\ (8.314 \text{ J K}^{-1} \text{ mol}^{-1})\ (300 \text{ K}) \times \log 10$$

$$\left\{\because \log \frac{V_2}{V_1} = \log 10\right\}$$

$$= (2.303)\ (8.314)\ (300 \text{ K})\ \log 10 \text{ J mol}^{-1}$$

$$= 5744 \text{ J mol}^{-1}$$

(b) When the gas obeys van der Waal's equation, work done by 1 mole of gas is

$$= 2.303 \text{ RT log } \frac{V_2 - b}{V_1 - b} + a\left\{\frac{1}{V_2} - \frac{1}{V_1}\right\}$$

After substituting the values of various quantities and solving, the work done is

$$= 5794 \text{ J mol}^{-1} - 252.4 \text{ J mol}^{-1}$$

$$= 5541.6 \text{ J mol}^{-1}$$

$$= 5.542 \text{ kJ mol}^{-1}.$$

Example 20:

One mole of a van der Walls's gas is allowed to expand isothermally and reversibly from a volume of 1 litre to 50 litre at 0°C. Calculate w, *q, Δ E and ΔH. van der Waal's constants are a = 6.5 atm l^2mole $^{-2}$ = 0.056 litre mole $^{-1}$ and R = 0.082 l atom deg $^{-1}$ mole $^{-1}$.*

Solution:

Here $V_1 = 1$ litre,

$V_2 = 50$ litres, n = 1 mole

T = 273 + 0 = 273°k,

R = 0.082 l-atm deg^{-1} $mole^{-1}$

a = 6.5 atm l^2 $mole^{-2}$,

b = 0.056 litre $mole^{-1}$

$$w = 2.303\ nRT \log\left(\frac{V_2 - nb}{V_1 - nb}\right) + an^2\left(\frac{1}{V_2} - \frac{1}{V_1}\right)$$

$$= 2.303 \times 1 \times 0.082 + 273\ [\log(50 - 0.056) - \log(1 - 0.056)) + 6.5\left(\frac{1}{50} - \frac{1}{1}\right)$$

= 82.47 l atm.

$$\Delta E = -\ an^2\left(\frac{1}{V_2} - \frac{1}{V_1}\right) = 6.5 \times \frac{49}{50}$$

$$q = 2.303nRT \log \frac{V_2 - b}{V_1 - b}$$

= ΔE + w = 884 l atm.

Example 21:

The van der Waal's constant a and b for hydrogen in litre atmosphere units are 0.246 and 2.67 × 10^{-2} respectively. Calculate the inversion temperature of hydrogen.

Solution:

we know T_i = 2a/Rb

or $$T_i = \frac{2 \times 0.246}{0.0821 \times 0.0267}$$

= 224.5°k = (224.5 – 273) = – 48.5°C.

Example 22:

How much heat is required to raise the temperature of 1 mole oxygen from 300K to 1300 K at constant pressure.

$$C_p = 6.095 + 3.325 \times 10^{-3}\ T - 1.017 \times 10^{-6}\ T^2.$$

Solution:

$$H_2 - H_1 \int_{T_t}^{T_2} C_p\ dT$$

$$T_1 = 300\ k,\ T_2 = 100\ k$$

$$\therefore H_{1300} - H_{300} = \int_{300}^{1300} \left(6.095 + 3.253 + 10^{-3}\ T - 1.017 \times 10^{-6}\ T^2\right)$$

[Substituting the value of C_p]

$$\because H_{1300} - H_{300} = 6.095\ (1300 - 300) + \frac{3.253 \times 10^{-3}}{2}$$

$$\times [(1300)^2 - (300)^2] - 1/3 \times 1.017 \times 10^{-6} [(1000)^2 - (300)^2]$$

$$= 6095 + 2602.4 - 735.63$$

$$= 7961.77 \text{ cal/mole.}$$

Example 23:

For equation (P _ a/V²) V = RT, prove that (i) dP is an exact differential, (ii) P is a state function and

$$\text{(iii)} \left(\frac{\partial P}{\partial T}\right)_V \left(\frac{\partial T}{\partial V}\right)_P \left(\frac{\partial V}{\partial T}\right)_T + 1 = 0.$$

Solution:

(i) For dP to be an exact differential, it is be proved that

$$\frac{\partial^2 P}{\partial V \partial T} = \frac{\partial^2 P}{\partial T \partial V} \quad ...(1)$$

$$(P + a/V^2)\ V = RT$$

or $$P = \frac{RT}{V} - \frac{a}{V^2} \quad ...(2)$$

When the above equation is differentiated with respect to V at constant temperature T, we obtain

$$\left(\frac{\partial P}{\partial V}\right)_T = -\frac{RT}{V^2} + \frac{2a}{V^3}$$

When the above equation is differentiated with respect to T at constant V, we obtain

$$\frac{\partial^2 P}{\partial T \partial V} = -\frac{RT}{V^2} \quad ...(3)$$

When equation (2) is differentiated with respect to T at constant volume V, we obtain

$$\left(\frac{\partial P}{\partial T}\right)_v = \frac{R}{V}$$

When the above equation is further differentiated with respect to V at constant T, we obtain

$$\frac{\partial^2 P}{\partial T \partial V} = -\frac{R}{V^2} \qquad ...(4)$$

From equations (3) and (4), we get

$$\frac{\partial^2 P}{\partial T\,\partial V} = \frac{\partial^2 P}{\partial T\,\partial V} = -\frac{R}{V^2} \qquad ...(5)$$

Thus, we have proved condition, (1). This shows that dP is an exact differential.

(ii) In the said question, first of all a change occurs in volume at constant temperature followed by a change in temperature at constant volume, thereby giving the final equation (3).

Again, a change in temperature is considered at constant volume, followed by a change in volume at constant temperature, thereby giving the final equation (4). Both equations (3) and (4) are same, thereby revealing that the final change in P remains the same irrespective of the difference in intermediate changes. Therefore, P is a state function as changes in P have been found to depend upon the intial and final states of the system.

(iii) Equation (2) is as follows :

$$P = \frac{RT}{V} - \frac{a}{V^2}$$

When the above equation is differentiated, we obtain :

$$dP = \frac{R}{V}\,dT - \frac{RT}{V^2}\,dV + \frac{2a}{V^3}\,dV$$

$$= \frac{R}{V}\,dT - \left(\frac{RT}{V^2} - \frac{2a}{V^3}\right)dV \qquad ...(6)$$

If volume is constant, dV = 0 : equation (6) becomes as follows :

$$dP = \frac{R}{V}\,dT \text{ or } \left(\frac{\partial P}{\partial V}\right)_v = \frac{R}{V} \qquad ...(7)$$

If pressure is constant, dP = 0; equation (6) becomes as follow :

$$\frac{R}{V}\,dT - \left(\frac{RT}{V^2} - \frac{2a}{V^3}\right)dV = 0$$

or
$$\left(\frac{\partial T}{\partial V}\right)_P = \left(\frac{RT}{V^2} - \frac{2a}{V^3}\right)\Big/\frac{R}{V} \quad ...(8)$$

If temperature is constant, dT = 0; equation (6) becomes as follows:

$$\left(\frac{\partial T}{\partial V}\right)_T = -\frac{1}{\left(\frac{RT}{V^2} - \frac{2a}{V^3}\right)} \quad ...(9)$$

On multiplying equations (7), (8) and (9), we obtain

$$\left(\frac{\partial p}{\partial T}\right)_V \left(\frac{\partial T}{\partial V}\right)_P \left(\frac{\partial V}{\partial P}\right)_T = -\frac{R}{V} \times \frac{\left(\frac{RT}{V^2} - \frac{2a}{V^3}\right)}{\left(\frac{RT}{V^2} - \frac{2a}{V^3}\right)} \times \frac{1}{R} = -1$$

or
$$\left(\frac{\partial p}{\partial T}\right)_V \left(\frac{\partial T}{\partial V}\right)_P \left(\frac{\partial V}{\partial P}\right)_T + 1 = 0$$

This is cyclic rule.

Example 24:

For an ideal gas PV = RT, prove that dP is an exact differential.

Solution:

We know $PV = RT$

or
$$P = \frac{RT}{V} \quad \text{(R is a constant)} \quad ...(1)$$

On differentiating with respect to T, taking V constant, we obtain

$$\left(\frac{\partial P}{\partial T}\right)_P = \frac{RT}{V} \quad ...(2)$$

Again, on differentiating with respect to V, we obtain

$$\frac{\partial^2 P}{\partial V \partial T} = -\frac{R}{V^2} \quad ...(3)$$

On differentiating (1) with respect to volume, taking T constant, we obtain

$$\left(\frac{\partial P}{\partial V}\right)_T = -\frac{RT}{V^2} \quad ...(4)$$

Again, on differentiating above equation with respect to temperature taking V as constant, we obtain

$$\frac{\partial^2 P}{\partial T \partial V} = -\frac{R}{V^2} \quad ...(5)$$

From equations (3) and (5), we obtain

$$\frac{\partial^2 P}{\partial T \partial V} = \frac{\partial^2 P}{\partial T \partial V}$$

Thus dP is an exact differential.

Example 25:

For an ideal gas PV = RT, prove that DT is an exact differential.

Solution:

$$PV = RT \quad ...(1)$$

In order to prove that dT is a perfect differential, we have to prove that

$$\frac{\partial^2 T}{\partial V \partial P} = \frac{\partial^2 T}{\partial P \partial V}$$

On differentiating Eq. (1), we obtain

$$\left(\frac{\partial T}{\partial V}\right)_P = \frac{P}{R}; \text{i.e., } \frac{\partial^2 T}{\partial P \partial V} = \frac{1}{R}$$

Again $$\left(\frac{\partial T}{\partial P}\right)_V = \frac{V}{R}; \text{i.e., } \frac{\partial^2 T}{\partial V \partial P} = \frac{1}{R}$$

Therefore, $$\frac{\partial^2 T}{\partial V \partial P} = \frac{\partial^2 T}{\partial P \partial V}$$

Thus, dT is an exact differential.

Example 26:

For an ideal gas (PV = nRT) show that (1/T) is an integrating factor for dw = P dV.

Solution:

$$dw = P\,dV \quad ...(1)$$

$\because$ $$PV = nRT$$

$\therefore$ $$p\,dV + V\,dP = nR\,dT$$

and $$dV = \frac{nR}{P}dT - \frac{V}{P}dP \quad ...(2)$$

On substituting the expression for dV in equation (1), we obtain

$$dw = nR\ dT - V\ dP = nR\ dT - \frac{nRT}{P}dP$$

Now $$\frac{\partial}{\partial P}(nR)_T = 0$$

and $$\frac{\partial}{\partial T}\left(-\frac{nRT}{P}\right)_P = -\frac{nR}{P}$$

Hence, dw is inexact differential.

Suppose f = 1/T be an integrating factor. Then

$$f\ .\ dw = \frac{1}{T}nR\,dT - \frac{nR}{P}dP$$

$$\therefore \quad \frac{\partial}{\partial P}\left[\frac{nR}{T}\right]_T = 0$$

and $$\frac{\partial}{\partial T}\left[\frac{nR}{P}\right]_T = 0$$

Thus f. dw$\frac{nR}{T}$dT – $\frac{nR}{P}$dP is an exact differential while 1/T is an integrating factor.

Example 27:

Calculate the maximum work obtainable by (i) the isothermal expansion and (ii) the adiabatic expansion of 2 mol of on ideal gas initially at 25° C form 10 L to 20 L. Assume Cv = (5/2) R.

Solution:

(i) Maximum work is obtained in reversible expansion. Hence

$$w = -nRT\ln\frac{V_2}{V_1}$$

$$= -(2\text{ mol})(8.314\text{ J K}^{-1}\text{ mol}^{-1})(298\text{ K})\left(2.303\log\frac{20L}{10L}\right.$$

= – 3435 J

(ii) Firstly, we calculate the final temperature using the expression

$$T_2 V_2^{\gamma-1} = T_1 V_1^{\gamma-1}$$

where $\gamma = \frac{C_P}{C_V} = \frac{(7/2)R}{(5/2)R} = \frac{7}{5}$

Hence, $T_2 = (298K)\left(\frac{10L}{20L}\right)^{(7/5)-1}$

$= (298\ K)\ (0.7578) = 225.8\ K$

Now $W_{adi} = \Delta E$

$= n\ C_{V},\ m(T_2 - T_1)$

$= (2mol)(2.5 \times 8.314\ JK^{-1}mol^{-1})(225.8K - 298K)$

$= -\ 3001\ J.$

Example 28:

Show that

$$\left(\frac{\partial \alpha}{\partial P}\right)_T + \left(\frac{\partial \beta}{\partial T}\right)_P = 0$$

where α and β are coefficient of thermal expansion and coefficient of compressibility, respectively.

Solution:

Since V is a state function, we can write

$$\frac{\partial^2 V}{\partial T\, \partial P} = \frac{\partial^2 V}{\partial P\, \partial T}$$

or $\frac{\partial}{\partial T}(-V\beta)_P = \frac{\partial}{\partial T}(V\alpha)_T$

Carrying out the differentiation, we get

$$-\beta\left(\frac{\partial V}{\partial T}\right)_P - V\left(\frac{\partial \beta}{\partial T}\right)_P = \alpha\left(\frac{\partial V}{\partial P}\right)_T + V\left(\frac{\partial \alpha}{\partial P}\right)_T$$

that is $\left(\frac{\partial \alpha}{\partial P}\right)_T + \left(\frac{\partial \beta}{\partial T}\right)_P = -\frac{\beta}{V}\left(\frac{\partial V}{\partial T}\right)_P - \frac{\alpha}{V}\left(\frac{\partial V}{\partial P}\right)_T$

$$= -\frac{\beta}{V}(\alpha V) - \frac{\alpha}{V}(-\beta V) = 0$$

Example 29:

Show that Joule Thomson coefficient, $\mu_{JT} = 0$ for an ideal gas. Comment on the liquefaction of an ideal gas.

Solution:

By definition, Joule-Thomson coefficient is defined as

$$\mu_{JT} = \left(\frac{\partial T}{\partial P}\right)_H$$

Using the cyclic rule

$$\left(\frac{\partial T}{\partial P}\right)_H \left(\frac{\partial P}{\partial H}\right)_T \left(\frac{\partial H}{\partial T}\right)_P + 1 = 0$$

we get $$\left(\frac{\partial T}{\partial P}\right)_H = -\frac{1}{\left(\frac{\partial P}{\partial H}\right)_T \left(\frac{\partial H}{\partial T}\right)_P} = -\frac{\left(\frac{\partial H}{\partial P}\right)_T}{C_P}$$

Now using, thermodynamic equation of state

$$\left(\frac{\partial T}{\partial P}\right)_H = V - T\left(\frac{\partial V}{\partial T}\right)_P$$

we get $$\mu_{JT} = \frac{V - T\left(\frac{\partial V}{\partial T}\right)_P}{C_P}$$

Now, for an ideal gas

$$V = \frac{nRT}{P}$$

Hence, $$\left(\frac{\partial V}{\partial T}\right)_P = \frac{nR}{P} = \frac{V}{T}$$

Thus $$\mu_{JT} = \frac{V - T\left(\frac{V}{T}\right)}{C_P} = 0$$

The ideal gas cannot be liquefied as there will not occur any change in temperature when the gas undergoes Joule-Thomson experiment.

Example 30:

Calculate the work down when 50 g of iron dissolves in hydrochloric acid at 25oC in :

(i) a closed vessel and

(ii) an open beaker.

(iii) In which case would maximum amount of work be done ?

Solution:

$$\text{Amount of iron} = \frac{50\text{ g}}{55.85\text{ g mol}^{-1}} = \frac{50}{55.85}\text{ mol}$$

(i) In a closed vessel, $\Delta V = 0$. Hence,

$$dw = -PdV = 0$$

that is, no work is involved when iron dissolves in hydrochloric acid in a closed vessel.

(ii) The reaction between iron and hydrochloric acid is

$$Fe + 2HCl \rightarrow FeCl_2 + H_2(g)$$

Thus, the liberated hydrogen gas pushes the atmospheric air and thus involves the work of expansion. In this case, we will have

$$w = -P_{ext}\ \Delta V$$
$$= -P_{ext}\ V_{H2}$$

where V_{H2} is the volume of hydrogen released. According to the ideal gas law, we will have

$$w = -P_{ext}\ (n_{H2}RT/P_{ext})$$
$$= -n_{H2}RT$$
$$= -(50/55.85)\text{ mol }(8.314\text{ J K}^{-1}\text{ mol}^{-1})\ (298\text{ K})$$
$$= -2.2\text{ kJ}$$

Example 31:

Since CV = (∂q/∂T) V by definition, one often writes without restriction $dV = C_V dT$. This is generally not true. Explain why.

Solution:

If we take U = f)V, T), then we will have

$$dU = \left(\frac{\partial U}{\partial V}\right)_T dV + \left(\frac{\partial U}{\partial T}\right)_V dT$$
$$= \left(\frac{\partial U}{\partial V}\right)_T \cdot dV + C_V\ dT$$

Thus, if one writes $dV = C_V\ dT$ without restriction then this statement is not complete. This is applicable for a system in which $(\partial U/\partial V)_T = 0$. An ideal gas is such an example. Alternatively, the system may be undergoing change at constant volume, for which, $(\partial U/\partial V)_T\, dV$ is zero and thus $dV = C_V\, dT$.

Example 32:

Calculate the work done by the system when 2 mol of an ideal gas expand from 0.01 m³ to 0.1 m³ at 300 k isothermally :

(i) into a vacuum, and

(ii) reversibly.

Explain how the surroundings can be restored only in one case and not in the other.

Solution:

(i) When the work is done into a vacuum, the work done by the system is zero since P_{opp} in this case is zero.

(ii) For the reversible isothermal expansion work involved is

$$w = -\, nRT \ln \frac{V_2}{V_1}$$

$$= (2)\ (8.314 \times 300\ J)\ \ln \frac{0.1}{0.01}$$

$$= -\, 2 \times 8.314 \times 300 \times 2.303\ J$$

$$= -\, 11488\ J = -\, 11.488\ kJ.$$

This surroundings are not disturbed when the expansion is done against vacuum. In case of reversible isothermal expansion, heat is absorbed by the system from the surroundings. Now if we carry out reversible isothermal compression of the system till the original state, then the same quantity or heat is released to the surroundings and thus the latter is restored back to the original state. However, in the first case where expansion is done against vacuum, there will be bet heat received by the surroundings during compression. Hence, in this case, it is not restored back to its original state.

Example 33:

Show that PV = nRt, then

(i) $$\frac{\partial^2 P}{\partial n \partial V} = \frac{\partial^2 P}{\partial V \partial n} = -\frac{RT}{V^2}$$

(ii) $$\frac{\partial^2 P}{\partial n \partial T} = \frac{\partial^2 P}{\partial T \partial n} = -\frac{R}{V}$$

(iii) $$\left(\frac{\partial P}{\partial T}\right)_V \left(\frac{\partial T}{\partial V}\right)_P \left(\frac{\partial V}{\partial P}\right)_T = -1$$

Solution:

(i) $\because$ $PV = nRT$

$\therefore$ $P = \frac{nRT}{V}$

or $$\frac{\partial P}{\partial V} = \frac{\partial}{\partial V}\left(\frac{nRT}{V}\right)$$ [keeping T as constant]

$$= nRT \frac{\partial}{\partial V}\left(\frac{1}{V}\right) = -\frac{nRT}{V^2}$$

$$\frac{\partial^2 P}{\partial n \partial V} = \frac{\partial}{\partial n}\left(\frac{\partial P}{\partial V}\right)$$

$$= \left(-\frac{nRT}{V^2}\right)$$ [Keeping R, T and V as constant]

$$= -\frac{RT}{V^2} \quad ...(1)$$

Similarly, $PV = nRT$ or $P = \frac{nRT}{V}$

$$\frac{\partial P}{\partial n} = \frac{RT}{V}$$

$$\frac{\partial^2 P}{\partial V \partial N} = \frac{\partial}{\partial V}\left(\frac{RT}{V}\right)$$ [keeping T and R as constant]

$$= -\frac{RT}{V^2} \quad ...(2)$$

From equations (1) and (2), we obtain

$$\frac{\partial^2 P}{\partial n \partial V} = \frac{\partial^2 P}{\partial V \partial n} = -\frac{RT}{V^2}$$

(ii) $$P = \frac{nRT}{V}$$

$$\frac{\partial P}{\partial T} = \frac{\partial}{\partial T}\left(\frac{nRT}{V}\right)$$ [Keeping n and R as constant]

$$= \frac{nR}{V}$$

$$\frac{\partial^2 P}{\partial n \partial T} = \frac{\partial}{\partial n}\left(\frac{nR}{V}\right) \text{[Keeping V and R as constant]}$$

$$= \frac{R}{V} \qquad ...(3)$$

Again, $P = \frac{nRT}{V}$

$$\frac{\partial P}{\partial n} = \frac{RT}{V} \qquad \text{[Keeping RT and V as constant]}$$

$$\frac{\partial^2 P}{\partial T \partial n} = \frac{\partial}{\partial n}\left(\frac{RT}{V}\right) \qquad \text{[Keeping RT and V as constant]}$$

or $\frac{\partial^2 P}{\partial T \partial n} = \frac{R}{V}$ ...(4)

From equations (3) and (4), we obtain

$$\frac{\partial^2 P}{\partial n \partial T} = \frac{\partial^2 P}{\partial T \partial n} = \frac{R}{V}$$

(iii) Again, $P = \frac{nRT}{V}$

$$\left(\frac{\partial P}{\partial T}\right)_v = \frac{nR}{V} \qquad ...(5)$$

Again, $V = \frac{nRT}{P}$

$$\left(\frac{\partial V}{\partial P}\right)_T = -\frac{nRT}{P} \qquad ...(6)$$

Again, $T = \frac{PV}{nR}$

$$\left(\frac{\partial T}{\partial V}\right)_P = -\frac{P}{nR} \qquad ...(7)$$

On multiplying equations (5), (6), and (7), we obtain

$$\left(\frac{\partial P}{\partial T}\right)_V \left(\frac{\partial T}{\partial V}\right)_P \left(\frac{\partial V}{\partial P}\right)_T = \frac{nR}{V} \times -\frac{nRT}{P^2} \times \frac{P}{nR}$$

$$= -\frac{nRT}{PV} = -1\left[\because 1 = \frac{nRT}{PV}\right]$$

Example 34:

(i) Calculate ∝ and β for an ideal gas obeying gas equation PV = RT (for one mole).

Solution:

(i) We know

$$PV = RT$$

On differentiating the above equation, we obtain

$$PdV + VdP = RdT \qquad ...(1)$$

At constant pressure dP = 0, the above equation becomes as follows:

$$\left(\frac{\partial V}{\partial T}\right)_P = \frac{R}{V}$$

or $$\alpha = \frac{1}{V}\left(\frac{\partial V}{\partial T}\right)_P = \frac{R}{VP} = \frac{1}{T} = T^{-1}$$

At constant temperature dT = 0; equation (1) becomes as follows:

$$\left(\frac{\partial V}{\partial P}\right)_T = -\frac{V}{P}$$

or $$\beta = -\frac{1}{V}\left(\frac{\partial V}{\partial P}\right)_T = \frac{V}{VP} = \frac{1}{P} = P^{-1}$$

(ii) The van der Waal's equation is as follows :

$$\left(P + \frac{a}{V^2}\right)(v - b) = RT$$

or $$PV^3 - PbV^2 + aV - ab - RTV^2 = 0 \qquad ...(2)$$

On differentiating the above equation with respect to T at constant pressure, we obtain

$$3PV^2\left(\frac{\partial V}{\partial T}\right)_P - 2PbV\left(\frac{\partial V}{\partial T}\right)_P + a\left(\frac{\partial V}{\partial T}\right)_P - 2RTV\left(\frac{\partial V}{\partial T}\right)_P - RV^2 = 0$$

or $$\left(\frac{\partial V}{\partial T}\right)_P = \frac{RV^2}{3PV^2 - 2PbV + a - 2RTV}$$

$$\alpha = \frac{1}{V}\left(\frac{\partial V}{\partial T}\right)_P$$

$$= \frac{1}{V}\left[\frac{RV^2}{3PV^2 - 2PVb + a - 2RTV}\right]$$

$$= \frac{R}{3PV - 2Pb + \frac{a}{V} - 2RT}$$

But the van der Waal's equation may be put as follows :

$$RT = PV - Pb + \frac{a}{V} - \frac{ab}{V^2}$$

$$\therefore \qquad \alpha = \frac{R}{3PV - 2Pb + \frac{a}{V} - 2\left(PV - Pb + \frac{a}{V} - \frac{ab}{V^2}\right)}$$

$$= R\left[PV - \frac{a}{V} + \frac{2ab}{V^2}\right]^{-1}$$

Again, on differentiating equation (2), with respect to P at constant temperature T, we obtain

$$3PV^2\left(\frac{\partial V}{\partial P}\right)_T + V - 2PbV\left(\frac{\partial V}{\partial P}\right)_T$$

$$- bV^2 + a\left(\frac{\partial V}{\partial P}\right)_T - 2RTV\left(\frac{\partial V}{\partial P}\right)_T = 0$$

$$\left(\frac{\partial V}{\partial P}\right)_T = \frac{V^2(b - V)}{3PV^2 - 2PbV + a - 2RTV}$$

But the van der Waal's equation may be put as follows :

$$RTV = PV^2 + a - PbV - ab/V$$

$$\therefore \left(\frac{\partial V}{\partial P}\right)_T = \frac{V^2(b - V)}{3PV^2 - 2PbV + a - 2\left(PV^2 + a - PbV - \frac{ab}{V}\right)}$$

$$= \frac{b - V}{P - \frac{a}{V^2} + \frac{2ab}{V^3}}$$

Now $\quad \beta = -\frac{1}{V}\left(\frac{\partial V}{\partial P}\right)_T = -\frac{b - V}{PV - \frac{a}{V} + \frac{2ab}{V^2}}$

$$= -\frac{V - b}{P - \frac{a}{V} + \frac{2ab}{V^2}}$$

Example 35:

Derive $C_{P,m} - C_{V,m} = TV_m \frac{\alpha^2}{\beta}$

where α *and* β *are the coefficient of thermal expansion and coefficient of compressibility, respectively.*

Solution:

We start with H = f (T, P) and U = f (T, V) and write their differentials as

$$dH = \left(\frac{\partial H}{\partial T}\right)_P dT + \left(\frac{\partial H}{\partial P}\right)_T dP$$

$$dU = \left(\frac{\partial U}{\partial T}\right)_V dT + \left(\frac{\partial U}{\partial V}\right)_T dV$$

Now since H = U + PV

we have dH = dU + P dV + V dP

$$\text{or} \left(\frac{\partial H}{\partial T}\right)_P dT + \left(\frac{\partial H}{\partial P}\right)_T dP = \left(\frac{\partial U}{\partial T}\right)_V dT + \left(\frac{\partial U}{\partial V}\right)_T dV + P\,dV + V\,dP$$

Dividing by dT, keeping P constant, we get

$$\left(\frac{\partial H}{\partial T}\right)_P = \left(\frac{\partial U}{\partial T}\right)_V + \left(\frac{\partial U}{\partial V}\right)_T \left(\frac{\partial V}{\partial T}\right)_P + P\left(\frac{\partial V}{\partial T}\right)_P$$

Now using the thermodynamic equation of state

$$\left(\frac{\partial U}{\partial V}\right)_T = T\left(\frac{\partial P}{\partial T}\right)_V - P$$

we get $$\left(\frac{\partial H}{\partial T}\right)_P = \left(\frac{\partial U}{\partial T}\right)_V + T\left(\frac{\partial P}{\partial T}\right)_V \left(\frac{\partial V}{\partial T}\right)_P \quad ...(1)$$

Now, by definition

$$\alpha = \frac{1}{V}\left(\frac{\partial V}{\partial T}\right)_P \quad ...(2)$$

$$\beta = -\frac{1}{V}\left(\frac{\partial V}{\partial P}\right)_T \quad ...(3)$$

Using the cyclic rule

$$\left(\frac{\partial V}{\partial P}\right)_T \left(\frac{\partial P}{\partial T}\right)_V \left(\frac{\partial T}{\partial V}\right)_P + 1 = 0$$

we get $$\left(\frac{\partial V}{\partial P}\right)_T = -\left(\frac{\partial V}{\partial T}\right)_P \left(\frac{\partial T}{\partial P}\right)_V$$

Hence

$$\beta = \frac{1}{V}\left(\frac{\partial V}{\partial T}\right)_P \cdot \left(\frac{\partial T}{\partial P}\right)_V$$

and $$\frac{\alpha}{\beta} = \left(\frac{\partial P}{\partial T}\right)_V$$

Hence, Eq (1) becomes

$$C_{P,m} - C_{V,m} = T\left(\frac{\partial P}{\partial T}\right)_V \left(\frac{\partial V}{\partial T}\right)_P \qquad ...(4)$$

With the use of Eqs. (2) and (4), we get

$$C_{P,m} - C_{V,m} = TV\,\frac{\alpha^2}{\beta}$$

Example 36:

When a bicycle tyre is inflated with hand pump, the temperature of the air inside increases. Explain.

Solution:

Work done on the system is converted into heat which increases the temperature of air inside the tyre.

Example 37:

Can q become a state function ?

Solution:

Heat exchange reversibly at constant volume (or at constant pressure) is equal to the change in internal energy (or enthalpy) of a system. Since the latter is a state function, it follows that in such conditions q also behaves like a state function.

Example 38:

One mole of hydrogen and nine moles of nitrogen are mixed at 291. K and I atm pressure. Assuming ideal behaviour for the gases, calculate the entropy of mixing per mole of the mixture formed. Would it make any difference if under similar conditions one mole of hydrogen is mixed with nine moles of oxygen.

Solution:

For ideal gases the entropy of mixing per mole of the mixture is given as follows

$$\Delta S_{mtxing} = -R\sum X_i \ln X_i$$

In this case $X_{H2} = 0.1$ and $X_{N2} = 0.9$

$$\Delta S mixing = -R\,[0.1 \ln 0.1 + 0.9 \ln 0.9]$$

$$= -(8.314\ JK^{-1}\ mole^{-1})\ (2.303)\ (0.1)$$

$$[\because \quad R = 8.314\ \ JK^{-1}\ mol^{-1}]$$

$$\left[\log\frac{1}{10} + 9\log\frac{9}{10}\right]$$

$$= 2.704 \quad JK^{-1}\ mol^{-1}$$

When hydrogen is mixed with oxygen, the entropy of mixing would remain unchanged provided oxygen also behaves ideally.

Example 39:

The temperature of an ideal monoatomic gas is increased from 546 K to 1638 K. Calculate the pressure change in order that the entropy of the gas remains unchanged in the process

Solution:

The entropy change for the system can be explained by the following relation which is as follows

$$\Delta S_{sys} = nC_P \ln\frac{T_2}{T_1}\ nR \ln\frac{P_2}{P_1}$$

$$\Delta S_{sys} = 0, \text{ thus}$$

$$C_P \ln\frac{T_2}{T_1} = R \ln\frac{P_2}{P_1}$$

$$\frac{5}{2} R \log\frac{T_2}{T_1} = R \log\frac{P_2}{P_1}$$

$$\frac{5}{2} \log\frac{1638\ K}{546\ K} = \log\frac{P_2}{P_1}$$

$$\frac{P_2}{P_1} = 15.60$$

Thus in order to make entropy constant, the pressure of the gas must be increased by 15.60 times its initial value.

Example 40:

2 moles of an ideal monoatomic gas (C_V = 20.91 J mol^{-1} K^{-1}) are heated from 700 K to 750 K. The volume of the gas changes from 0.08 m^3 to 0.8 m^3. Assuming that the heat capacity remains constant in this temperature range, calculate the entropy change for the system, the surroundings and the universe if the process is carried out (a) reversibly and (b) irreversibly by placing the system in contact with a reservoir at 500 K and allowing the gas to expand against a const. external pressure equal to the final pressure of the gas

Solution:

For the reversible transformation of an ideal gas the entropy change of the system is given as follows :

$$\Delta S_{system} = nC_V \ln \frac{T_2}{T_1} + nR \ln \frac{V_2}{V_1}$$

$$= (2 \text{ mol}) (20.9 \text{ J mol}^{-1} \text{ K}^{-1}) (2.303) \log \frac{750}{700})$$

$$+ (2 \text{ mol}) (8.314 \text{ J mol}^{-1} \text{ K}^{-1}) (2.303) \log \frac{0.80}{0.08}$$

$$= 6.449 + 38.28 = 44.729 \text{ JK}^{-1}.$$

As the process is reversible, the entropy change of the surrounding is as follows.

$$\Delta S_{sur} = - 44.729 \text{ JK}^{-1}$$

and the entropy change of the universe is as follows

$$\Delta S_{univ} = \Delta S_{sys} + \Delta S_{surr} = 0$$

(b) In the irreversible transformation the entropy change of the system will be the same. Heat lost by the surrounding is as

$$- Q = \Delta E - W$$

$$\Delta E = - n C_V (750 - 700)$$

and $$\omega = P_{ext} (V_2 - V_1).$$

Thus according to the final state

$$T_2 = 750 \text{ K}$$

$$V_2 = 0.8 \text{ m}^3$$

and hence the pressure is given at follows

$$P_2 = \frac{nRT_2}{V_2} = \frac{(2 \text{ mol})\left(8.314 \text{ J mol}^{-1} \text{ K}^{-1}\right)(750\text{K})}{\left(0.8 \text{ m}^3\right)}$$

$$= 15588.75 \text{ Jm}^{-3}.$$

Now $P_{ext} = P_2 = 15588.75 \text{ Jm}^{-3}$ and hence the heat lost by the surrounding is

$$-Q = -(2 \text{ mol})(20.9 \text{ J mol}^{-1} \text{ K}^{-1})(50 \text{ K})$$

$$-(15588.75 \text{ Jm}^{-3})(0.8 \text{ m}^3 - 0.08 \text{ m}^3)$$

$$= -2090 \text{ j} - 11223.9$$

$$= 13313.9 \text{ J}.$$

Let us assume that this heat is lost reversibly by the surroundings. Its entropy change is given as

$$\Delta S_{surr} = \frac{13313.9 \text{ j}}{500 \text{ K}} = 26.6278 \text{JK}^{-1}$$

and the entropy change of the universe is

$$\Delta S_{univ} = \Delta S_{sys} + \Delta S_{surr}$$

$$= (44.729 - 26.6278) \text{ JK}^{-1}$$

$$= 18.1012 \text{ JK}^{-1}.$$

Example 41:

Two moles of an ideal gas are allowed to expand isothermally from 0.04 m^3 to 0.4 m^3 at 300 K. Calculate the entropy change for the system, surroundings and universe if the expansion is

(i) reversible and

(ii) irreversible against a constant external pressure of 0.2 atm.

Solution:

$$Q_{rev} = W = n \text{ RT ln } \frac{V_2}{V_1}$$

$$\Delta S_{rev} = \frac{Q_{rev}}{T} = nR \ln \frac{V_2}{V_1}$$

$$= (2 \text{ mol}) (8.314) \text{ J mol}^{-1} \text{ K}^{-1} (2.303) \log \frac{0.40}{0.04}$$

$$= 38.294284 \text{ JK}^{-1}.$$

The surroundings is losing an exactly equivalent amount of heat and hence its entropy decrease is given as follows :

$$\Delta S_{surr} = \frac{(-Q_{rev})}{T} = -19.14 \text{ JK}^{-1}$$

The entropy change of the universe is given as follows :

$$\Delta S_{univ} = \Delta S_{sys} + \Delta S_{surr} = 0$$

(ii) In the irreversible expansion, the entropy change of the system is still the same as in the reversible change, *i.e.*, 19.14 JK^{-1} but heat lost by the surroundings is given by $-P\Delta V$ and consequently the entropy change, if the transfer were reversible, is

$$\Delta S_{surr} = \frac{Q_{surr}}{T} = -\frac{P\Delta V}{T}$$

$$= -\frac{(0.2 \text{ atm.})(0.4 - 0.04 \text{ m}^3)(101325 \text{ Jm}^{-3} \text{ atm}^{-1})}{300 \text{ K}}$$

$$= -\frac{0.2 \times 0.36 \times 101325}{300} \text{ JK}^{-1}$$

$$= 24.318 \text{ JK}^{-1}..$$

Example 42:

Calculate the entropy change when 1 mole of perfect gas is allowed to expand at 27°C from a volume of 2 litres to a volume of 20 litres against a constant pressure of one atmosphere.

Solution:

We know

$$dS = C_V \frac{dT}{T} + R \frac{dV}{V}$$

By integrating this equation, we get

$$S_2 - S_1 = C_V \ln \frac{T_2}{T_1} + R \ln \frac{V_2}{V_1}$$

At constant temperature,

$$S_2 - S_1 = R \ln \frac{V_2}{V_1}$$

$$= 2.303\ R \log \frac{V_2}{V_1}$$

$$= 2.303 \times 2 \times \log \frac{20}{2}$$

$$= 2.303 \times 2 \times \log 10$$

$$= 4.606 \text{ cal./ degree/mole.}$$

$$\left[\begin{array}{l} \because R = 2 \text{ cal./degree/mole} \\ V_1 = 2 \text{ litres} \\ V_2 = 20 \text{ litres} \end{array}\right]$$

Example 43:

The molar heat capacity at capacity at constant pressure of solid magnesium (0° – 600°C) is expressed by

CP = 6.20 + 1.30 × 10–3 T – 6.80 × 104 T–2

Calculate the increase in entropy when 1 gm atom of metal is heated from 27°C to 127°C at constant pressure.

Solution:

At constant pressure,

$$\Delta S = \int_{T_1}^{T_2} \frac{C_P}{T}\, dT$$

$$\therefore \quad \Delta S = \int_{300}^{400} \frac{\left(6.20 + 1.30 \times 10^{-3}\, T - 6.80 \times 10^4\, T^{-2}\right) dT}{T}$$

$$\int_{300}^{400} 6.20 \frac{dT}{T} + \int_{300}^{400} 1.3 \times 10^{-3} dT - \int_{300}^{400} 6.20 \times 10^4 \frac{dT}{T^3}$$

$$= 6.20 \ln\left[\ln T\right]_{300}^{400} + 1.3 \times 10^{-3} \left[T\right]_{300}^{400} - \frac{6.80 \times 10^4}{-2} \left[\frac{1}{T^2}\right]_{300}^{400}$$

$$= 6.20 \ln \frac{400}{300} + 1.3 \times 10^{-3} \times (400 - 300) + 3.40 \times 10^4$$

$$\left[\frac{1}{(400)^2} - \frac{1}{(300)^2}\right]$$

= 1.7487 cal./degree.

Example 44:

A thermostat was maintained at 370.05 K. For nearly an hour 4.2 kJ of heat leaked through the thermostat insulation into a room where the initial temperature of air was 300.05 K. (i) What was the entropy change of the material in the thermostat ? (ii) What was the entropy change of the air in the room ? (iii) Was the process spontaneous? (Material in the thermostat is water).

Solution:

(i) Entropy change of the material in the thermostat

$$-\frac{4.2 \text{ kJ}}{370.05\text{K}} = -11.35\times 10^3 \text{ kj K}^{-1}$$

(ii) Entropy change of the air in the room

$$= \frac{4.2 \text{ kJ}}{300.05 \text{ K}} = 14.00 \times 10^3 \text{ kJ K}^{-1}$$

(iii) Now

$$\Delta S_{total} = \Delta S_{thermostat} + \Delta S_{room}$$
$$= (-11.35 + 14.00) \text{ JK}^{-1}$$
$$= 2.65 \text{ J K}^{-1}$$

Since ΔStotal is positive, the process of leaking is spontaneous.

Example 45:

Calculate ΔS, ΔH and ΔU for the process

$$H_2 O(l, 20^oC, 1 \text{ atm} \rightarrow H_2O \text{ (g, } 250^oC, 1 \text{ atm)}$$

Given the following data :

$$C_P (l) = 75.6 \text{ J K}^{-1} \text{ mol}^{-1}$$

$$C_P (g) = 36.2 \text{ J K}^{1} \text{ mol}^{-1}$$

ΔH for vaporization of H_2O at 100°C and 1 atm is equal to 40.85 k J mol [1].

Solution:

The given process may be replaced by the following processes.

(i) H_2O (l, 20°C, 1 atm) → H_2O (l, 100°C, 1 atm)

(ii) H_2O (l, 100°C, 1 atm) → H_2O (g, 100°C, 1 atm)

(iii) H_2O (g, 100°C, 1 atm) → H_2O (g, 250°C, 1 atm)

Thus we have

$$\Delta S = \Delta S_{(1)} + \Delta S_{(ii)} \Delta S_{(iii)}$$

$$= C_{p,m}(l) \ln \frac{T_2}{T_1} + \frac{\Delta H_{vap}}{T} + C_{p,m}(g) \ln \frac{T_2'}{T_1'}$$

$$= \left(75.6 \ln \frac{373}{293} + \frac{40850}{373} + 36.2 \ln \frac{523}{373}\right) JK^{-1} \text{ mol}^{-1}$$

$$= (18.25 + 109.52 + 12.24) \text{ J K}^{-1} \text{ mol}^{-1}$$

$$= 140.01 \text{ J K}^{-1} \text{ mol}^{-1}$$

$$\Delta H = \Delta H_{(1)} + \Delta H_{(ii)} + \Delta H_{(iii)}$$

$$= C_{p,m} (l) (T_2 - T_1) + \Delta H_{vap} + C_{pm}(g) (T'_2 - T'_1)$$

$$= (75.6 \times 80 + 40850 + 36.2 \times 150) \text{ J mol}^{-1}$$

$$= 52328 \text{ J mol}^{-1}$$

$$\Delta U = \Delta U_{(1)} + \Delta U_{(ii)} \Delta U_{(iii)}$$

$$= C_{v,m} (l) (T_2 - T_1) + \Delta U_{vap} + C_{v,m}(g) (T'_2 - T'_1)$$

$$= C_{p,m}(l)(T_2 - T_1) + \{\Delta H_{vap} + \Delta_{V\,g} RT\} + \{C_{p,m}(g) - R\}(T'_2 - T'_1)$$

$$= [75.6 \times 80 + \{40850 - (8.314)(373)\}$$

$$+ (36.2 - 8.314)(150)] \text{ J mol}^{-1}\}$$

$$= (6040 + 37749 + 4183) \text{ J mol}^{-1}$$

$$= 47972 \text{ J mol}^{-1}.$$

Example 46:

Describe a process through with the gas can be restored to its initial state in each of the above cases. Explain how the surroundings can be restored only in one ca: e and not in the other.

Solution:

The gas can be restored to its initial state by compressing it isothermally from 5 dm^3 to 1 dm^3. This may be done reversibly or irreversibly. When the compression is done reversibly, the surroundings in the case of (ii) will be restored whereas in case of (i), it cannot be restored whether the compressions is done reversibly or irreversibly.

This is because, while on expansion, surroundings remain unchanged whereas during compression, it has to do work on the system and thus suffers changes.

Example 47:

Derive an expression for the entropy change when a van der Waals gas is heated and expanded simultaneously.

Solution:

Taking S = f(T, V), we get

$$dF = \left(\frac{\partial S}{\partial T}\right)_V dT + \left(\frac{\partial S}{\partial V}\right)_T dV$$

$$= \frac{nC_{V,m}}{T} dT + \left(\frac{\partial P}{\partial T}\right)_V dV$$

Now for van der Waal's gas we have

$$P = \frac{n\,RT}{V-nb} - \frac{n^2 a}{V^2}$$

Therefore $\left(\frac{\partial P}{\partial T}\right)_V = \frac{n\,RT}{V-nb}$

Thus $dS = \frac{n\,C_{V,m}}{T} dT + \frac{n\,R}{V-nb} dV$

Which on integrating within the limit T_1, V_1 and T_2, V_2 gives

$$\Delta S\; nC_{V,m} \ln \frac{T_2}{T_1} + nR \ln \frac{V_2 - nb}{V_1 - nb}$$

Example 48:

For an adiabatic process, dq = 0, If one were to write

$$\Delta S = \int \frac{dq}{T} = \int \frac{0}{T} = 0$$

Then every adiabatic process would be iso-entropic process. Comment.

Solution:

For an adiabatic process, dq – 0, and according to the first law of thermodynamics, dq = dU – dw

Thus, we will have

$$dU = dw$$

The expression of entropy change will be given as

$$\Delta S = \int \frac{dU - dw}{T}$$

$$= \int \frac{dU}{T} + \int \frac{P_{opp}}{T} dV$$

Now, it is only for reversible process the first integral is equal and opposite sign of the second integral and thus $\Delta S = 0$. For an irreversible process, these two integrals do not have equal magnitude and thus sum is not zero. Hence $\Delta S \neq 0$ for an adiabatic irreversible process.

Example 49:

The molar heat capacity at capacity at constant pressure of solid magnesium (0° – 600°C) is expressed by

CP = 6.20 + 1.30 × 10 – 3 T – 6.80 × 104 T – 2

Calculate the increase in entropy when 1 gm atom of metal is heated from 27°C to 127°C at constant pressure.

Solution:

At constant pressure,

$$\Delta S = \int_{T_1}^{T_2} \frac{C_P}{T} dT$$

$$\therefore \quad \Delta S = \int_{300}^{400} \frac{\left(6.20 + 1.30 \times 10^{-3} T - 6.80 \times 10^4 T^{-2}\right) dT}{T}$$

$$\int_{300}^{400} 6.20 \frac{dT}{T} + \int_{300}^{400} 1.3 \times 10^{-3} dT - \int_{300}^{400} 6.20 \times 10^4 \frac{dT}{T^3}$$

$$= 6.20 \ln[\ln T]_{300}^{400} + 1.3 \times 10^{-3} [T]_{300}^{400} - \frac{6.80 \times 10^4}{-2}\left[\frac{1}{T^2}\right]_{300}^{400}$$

$$= 6.20 \ln \frac{400}{300} + 1.3 \times 10^{-3} \times (400 - 300) + 3.40 \times 10^4$$

$$\left[\frac{1}{(400)^2} - \frac{1}{(300)^2}\right]$$

$$= 1.7487 \text{ cal./degree.}$$

Example 50:

Calculate the increase in entropy of three moles of hydrogen as it changes from 300°K atm to 1000°K and 1 atm (C_P = 7 cal./degree/ mole).

Solution:

Entropy change ΔS for n moles

$$= n\, C_P \ln \frac{T_2}{T_1}\, n\, R \ln \frac{P_2}{P_1}$$

$$= 3 \times 7 \times 2.303 \log \frac{1000}{300} - 3 \times 2 \times 2.303 \log \frac{10}{0.1}$$

$$= 25.287996 - 13.818$$

$$= 11.469996 \text{ cal./degree.}$$

Example 51:

Calculate ΔS, ΔH and ΔU for the process

$$H_2O(l,\ 20°C,\ 1\ atm \rightarrow H_2O\ (g,\ 250°C,\ 1\ atm)$$

Given the following data :

$$C_P\ (l) = 75.6\ J\ K^{-1}\ mol^{-1}$$

$$C_P\ (g) = 36.2\ J\ K^{-1}\ mol^{-1}$$

ΔH for vaporization of H_2O at 100°C and 1 atm is equal to 40.85 k J mol^{-1}.

Solution:

The given process may be replaced by the following processes.

(i) H_2O (l, 20°C, 1 atm) → H_2O (l, 100°C, 1 atm)

(ii) H_2O (l, 100°C, 1 atm) → H_2O (g, 100°C, 1 atm)

(iii) H_2O (g, 100°C, 1 atm) → H_2O (g, 250°C, 1 atm)

Thus we have

$$\Delta S = \Delta S_{(1)} + \Delta S_{(ii)}\ \Delta S_{(iii)}$$

$$= C_{p,m}(1) \ln \frac{T_2}{T_1} + \frac{\Delta H_{vap}}{T} + C_{p,m}(g) \ln \frac{T_2'}{T_1'}$$

$$= \left(75.6 \ln \frac{373}{293} + \frac{40850}{373} + 36.2 \ln \frac{523}{373}\right) JK^{-1}\ mol^{-1}$$

$= (18.25 + 109.52 + 12.24)\ J\ K^{-1}\ mol^{-1}$

$= 140.01\ J\ K^{-1}\ mol^{-1}$

$\Delta H = \Delta H_{(1)} + \Delta H_{(ii)} + \Delta H_{(iii)}$

$= C_{p,m}\ (l)\ (T_2 - T_1) + \Delta H_{vap} + C_{pm}(g)\ (T'_2 - T'_1)$

$= (75.6 \times 80 + 40850 + 36.2 \times 150)\ J\ mol^{-1}$

$= 52328\ \ J\ mol^{-1}$

$\Delta U = \Delta U_{(1)} + \Delta U_{(ii)}\ \Delta U_{(iii)}$

$= C_{v,m}\ (l)\ (T_2 - T_1) + \Delta U_{vap} + C_{v,m}(g)\ (T'_2 - T'_1)$

$= C_{p,m}(l)(T_2 - T_1) + \{\Delta H_{vap} + \Delta_{V\,g}RT\} + \{C_{p,m}(g) - R\}(T'_2 - T'_1)$

$= [75.6 \times 80 + \{40850 - (8.314)(373)\}$

$+ (36.2 - 8.314)(150)]\ J\ mol^{-1}\}$

$= (6040 + 37749 + 4183)\ J\ mol^{-1}$

$= 47972\ J\ mol^{-1}$.

Example 52:

For N = 400, find the ratios of the probabilities for (300, 100) and (200, 200) distribution in two identical boxes.

Solution:

We have

$$P(300, 100) = \frac{400!}{300!\ 100!}\ \frac{1}{2^{400}}$$

$$P(200, 200) = \frac{400!}{200!\ 200!}\ \frac{1}{2^{400}}$$

$$\frac{P(300, 100)}{P(200, 200)} = \frac{200!\ 200}{300!\ 100!} \cong 10^{-5}$$

Example 53:

Eight similar coins are tossed, find the probability of three heads and t tails.

Solution:

We know $P(3, 5) = \frac{{}^8C_3}{2^8} = \frac{8\times7\times6}{3\times2\times1\times2^8}$

or $$P(3, 5) = \frac{7}{32}$$

Example 54:

A system consists of 6000 particles distributed in three energy states with equal spacing. The energy of the three states are $E_{l'} = 0$, $E_{2'} = x$ and $E_3 = 2x$. All the three slates have the same intrinsic probability g. At a certain instant there are 3000 particles in lower level, 2500 in the middle level and 500 in the upper level. Compare the relative probabilities with the distribution obtained, by the transfer of one particle from the middle to the lower level and one particle from the middle to the upper level and the original distribution.

Solution:

Let P_1 and P_2 be the probabilities in the two cases

In the first case,

$$N = 6000,$$

$$n_1 = 3000,$$

$$n_2 = 2500,$$

$$n_3 = 500.$$

$$P_1 = \frac{g^N}{n_1!n_2!n_3!}$$

$$P_1 = \frac{g^{6000}}{3000!2500!500!} \quad ...(i)$$

In the second case

$$N = 6000,$$

$$n_1 = 3001,$$

$$n_2 = 2498,$$

$$n_3 = 501.$$

$$P_1 = \frac{g^N}{n_1!n_2!n_3!}$$

$$P_1 = \frac{g^{6000}}{3001!2498!501!}$$

Dividing (ii) by (i)

$$\frac{P_2}{P_1} = \frac{3000!2500!500!}{3001!2498!501!}$$

$$\frac{P_2}{P_1} = \frac{2500 \times 2499!}{3001 \times 501!}$$

$$\frac{P_2}{P_1} = 4.157$$

$$\approx 4.2.$$

It means the transfer of one particle from the middle to the upper and one particle from the middle to the lower state has changed the probability by a factor 4.2. This shows that both these distributions are not near the equilibrium state.

Example 55:

A thermostat was maintained at 370.05 K. For nearly an hour 4.2 kJ of heat leaked through the thermostat insulation into a room where the initial temperature of air was 300.05 K.

(i) What was the entropy change of the material in the thermostat?

(ii) What was the entropy change of the air in the room ?

(iii) Was the process spontaneous? (Material in the thermostat is water).

Solution:

(i) Entropy change of the material in the thermostat

$$-\frac{4.2 \text{ kJ}}{370.05\text{K}} = -11.35 \times 10^3 \text{ kj K}^{-1}$$

(ii) Entropy change of the air in the room

$$= \frac{4.2 \text{ kJ}}{300.05 \text{ K}} = 14.00 \times 10^3 \text{ kJ K}^{-1}$$

(iii) Now

$$\Delta S_{total} = \Delta S_{thermostat} + \Delta S_{room}$$

$$= (-11.35 + 14.00) \text{ JK}^{-1}$$

$$= 2.65 \text{ J K}^{-1}$$

Since ΔStotal is positive, the process of leaking is spontaneous.

Example 56:

At low temperature, a nonideal gas follows the relation PV = RT – a/V where a = 0.3636 Nm^4 $mole^{-2}$. Calculate the work done by one mole of this gas in expanding form 0.224 litre to 22.4 litre at 400 K. Compare this result with the work done in the corresponding expansion of an ideal gas.

Solution:

For an ideal gas, reversible work done

$$= 2.303 \text{ n RT log } \frac{V_2}{V_1}$$

V_1 = 0.224 litres $mole^{-1}$ = 22.4 × 10^{-5} m^3 $mole^{-1}$

P_2 = 22.4 litres $mole^{-1}$ = 22.4 × 10^{-3} m^3 $mole^{-1}$

R = 8.314 J K^{-1} $mole^{-1}$

$$W_{rew} = (2.303)\ (1 \text{ mole})\ (8.314 \text{ J K}^{-1} \text{ mole}^{-1})\ (400 \text{ K}) \times \log \frac{22.4 \times 10^{-3}}{22.4 \times 10^{-5}}$$

$$= (2.303)\ (1 \text{ mole})\ (8.314 \text{ J K}^{-1} \text{ mole}^{-1})\ (400 \text{ K}) \log 100$$

$$= 15320 \text{ J mole}^{-1}$$

$$= 15.32 \text{ kJ mole}^{-1}.$$

For the non-ideal gas PV = RT – a/V

$$\text{Therefore, the work done} = \int_{V_1}^{V_2} P\,dV = \int_{V_1}^{V_2} \left\{ \frac{RT}{V} - \frac{a}{V_2} \right\} dV$$

$$= 2.303 \text{ RT log } \frac{V_2}{V_1} + a \left\{ \frac{1}{V_2} - \frac{1}{V_1} \right\}$$

The value of first part is from derived eq. (1) as 15.32 kJ $mole^{-1}$ and of another part can be derived as

$$0.3636 \text{ Nm}^4 \text{ mole}^{-2} \left\{ \frac{1}{22.4 \times 10^{-3} \text{ m}^3 \text{ mole}^{-1}} - \frac{1}{22.4 \times 10^{-5} \text{ m}^3 \text{ mole}^{-1}} \right\}$$

$$= 1.604 \text{ kJ mole}^{-1}.$$

Thus work done = 15.320 k $Joule^{-1}$ – 1.604 k $Joule^{-1}$

= 13.716 kJ $mole^{-1}$.

Example 57:

What will be the work done when 65.38 g of zinc dissolves in hydrochloric acid in case of an open beaker and closed beaker at 300K.

Solution:

The process of dissolution of zinc in hydrochloric acid can be given as

$$Zn(s) + 2HCl\ (aq) = ZnCl_2(aq) + H_2(g)$$

From this equation it is clear that for each gram atom of zinc dissolved we will obtain 1 mole of Hydrogen gas. As a result of liberated gas, on the surrounding atmosphere the work performed will be equal to $P\Delta V$. If the initial volume of the system is neglected in comparison to the total volume of the gas produced and the gas shows ideal behaviour, then the work done in the open beaker is

$$W = P\Delta V \approx PV_{H2}$$

$$= nH2\ RT \qquad [\because T = 300\ k\ R = 8.314\ Jk^{-1}\ mole^{-1}]$$

$$= (1\ mole)\ (8.314\ JK^{-1}\ mol^{-1})\ (300\ K)$$

$$= 2.4942\ kJ.$$

If the reaction takes place in a closed beaker there is change in the volume *i.e.*, $\Delta V = 0$ hence work done is zero.

Example 58:

Calculate the work done when one mole of sulphur dioxide gas expands isothermally and reversibly at 300 k from $2.46 \times 10^{-3}\ m^3$ to $24.6 \times 10^{-3}\ m^3$. Assuming that the gas obeys

(a) ideal gas equation and

(b) van der Waals equation

$$\left\{ \begin{array}{l} a = 0.6799\ nm^4\ mole^{-2} \\ b = 0.564 \times 10^{-3}\ mole^{-1} \end{array} \right\}$$

Solution:

(a) The reversible work done by 1 mole of gas when it expands form $2.64 \times 10^{-3} m^3$ to $24.6 \times 10^{-3} m^3 = 2.303\ RT \log \frac{V_2}{V_1}$

$$= (2.303)\ (1\ mole)\ (8.314\ J\ K^{-1}\ mol^{-1})\ (300\ K) \times \log 10$$

$$\left\{\because \log\frac{V_2}{V_1} = \log 10\right\}$$

$$= (2.303)\ (8.314)\ (300\ K)\ \log 10\ J\ mol^{-1}$$

$$= 5744\ J\ mol^{-1}$$

(b) When the gas obeys van der Waal's equation, work done by 1 mole of gas is

$$= 2.303\ RT \log \frac{V_2 - b}{V_1 - b} + a\left\{\frac{1}{V_2} - \frac{1}{V_1}\right\}$$

After substituting the values of various quantities and solving, the work done is

$$= 5794\ J\ mol^{-1} - 252.4\ J\ mol^{-1}$$

$$= 5541.6\ J\ mol^{-1}$$

$$= 5.542\ kJ\ mol^{-1}.$$

Example 59:

One mole of a van der Walls's gas is allowed to expand isothermally and reversibly from a volume of 1 litre to 50 litre at 0°C. Calculate w, q, Δ E and Δ H. van der Waal's constants are a = 6.5atm $l^2 mole^{-2}$ = 0.056 litre mole $^{-1}$ and R = 0.082 l atom deg^{-1} $mole^{-1}$.

Solution:

Here $V_1 = 1$ litre,

$V_2 = 50$ litres, n = 1 mole

T = 273 + 0 = 273°k,

R = 0.082 l-atm deg^{-1} $mole^{-1}$

a = 6.5 atm l^2 $mole^{-2}$,

b = 0.056 litre $mole^{-1}$

$$w = 2.303\ nRT \log\left(\frac{V_2 - nb}{V_1 - nb}\right) + an^2\left(\frac{1}{V_2} - \frac{1}{V_1}\right)$$

$$= 2.303 \times 1 \times 0.082 + 273\ [\log(50 - 0.056) - \log(1 - 0.056)$$

$$+ 6.5\left(\frac{1}{50} - \frac{1}{1}\right)$$

$= 82.47 \text{ l atm.}$

$$\Delta E = -an^2\left(\frac{1}{V_2}-\frac{1}{V_1}\right) = 6.5 \times \frac{49}{50}$$

$$q = 2.303nRT \log \frac{V_2 - b}{V_1 - b}$$

$= \Delta E + w = 884 \text{ l atm.}$

Example 60:

The van der Waal's constant a and b for hydrogen in litre atmosphere units are 0.246 and 2.67 × 10^{-2} respectively. Calculate the inversion temperature of hydrogen.

Solution:

we know $T_i = 2a/Rb$

or $$T_i = \frac{2 \times 0.246}{0.0821 \times 0.0267}$$

$= 224.5°k = (224.5 - 273) = -48.5°C.$

Example 61:

How much heat is required to raise the temperature of 1 mole oxygen from 300K to 1300 K at constant pressure.

$$C_p = 6.095 + 3.325 \times 10^{-3}\,T - 1.017 \times 10^{-6}\,T^2.$$

Solution:

$$H_2 - H_1 \int_{T_t}^{T_2} C_p \, dT$$

$T_1 = 300$ k, $T_2 = 100$ k

$$\therefore H_{1300} - H_{300} = \int_{300}^{1300}\left(6.095 + 3.253 + 10^{-3}\ T - 1.017 \times 10^{-6}\ T^2\right)$$

[Substituting the value of C_p]

$$\because H_{1300} - H_{300} = 6.095\,(1300 - 300) + \frac{3.253 \times 10^{-3}}{2}$$

$$\times [(1300)^2 - (300)^2] - 1/3 \times 1.017 \times 10^{-6} [(1000)^2 - (300)^2]$$

$= 6095 + 2602.4 - 735.63$

$= 7961.77$ cal/mole.

Example 62:

For equation (P _ a/V²) V = RT, prove that (i) dP is an exact differential, (ii) P is a state function and

$$\text{(iii)} \left(\frac{\partial P}{\partial T}\right)_V \left(\frac{\partial T}{\partial V}\right)_P \left(\frac{\partial V}{\partial T}\right)_T + 1 = 0.$$

Solution:

(i) For dP to be an exact differential, it is be proved that

$$\frac{\partial^2 P}{\partial V \partial T} = \frac{\partial^2 P}{\partial T \partial V} \quad ...(1)$$

$$(P + a/V^2)\, V = RT$$

or
$$P = \frac{RT}{V} - \frac{a}{V^2} \quad ...(2)$$

When the above equation is differentiated with respect to V at constant temperature T, we obtain

$$\left(\frac{\partial P}{\partial V}\right)_T = -\frac{RT}{V^2} + \frac{2a}{V^3}$$

When the above equation is differentiated with respect to T at constant V, we obtain

$$\frac{\partial^2 P}{\partial T \partial V} = -\frac{RT}{V^2} \quad ...(3)$$

When equation (2) is differentiated with respect to T at constant volume V, we obtain

$$\left(\frac{\partial P}{\partial T}\right)_V = \frac{R}{V}$$

When the above equation is further differentiated with respect to V at constant T, we obtain

$$\frac{\partial^2 P}{\partial T \partial V} = -\frac{R}{V^2} \quad ...(4)$$

From equations (3) and (4), we get

$$\frac{\partial^2 P}{\partial T\, \partial V} = \frac{\partial^2 P}{\partial T\, \partial V} = -\frac{R}{V^2} \quad ...(5)$$

Thus, we have proved condition, (1). This shows that dP is an exact differential.

(ii) In the said question, first of all a change occurs in volume at constant temperature followed by a change in temperature at constant volume, thereby giving the final equation (3).

Again, a change in temperature is considered at constant volume, followed by a change in volume at constant temperature, thereby giving the final equation (4). Both equations (3) and (4) are same, thereby revealing that the final change in P remains the same irrespective of the difference in intermediate changes. Therefore, P is a state function as changes in P have been found to depend upon the intial and final states of the system.

(iii) Equation (2) is as follows :

$$P = \frac{RT}{V} - \frac{a}{V^2}$$

When the above equation is differentiated, we obtain :

$$dP = \frac{R}{V}dT - \frac{RT}{V^2}dV + \frac{2a}{V^3}dV$$

$$= \frac{R}{V}dT - \left(\frac{RT}{V^2} - \frac{2a}{V^3}\right)dV \qquad ...(6)$$

If volume is constant, dV = 0 : equation (6) becomes as follows :

$$dP = \frac{R}{V}dT \text{ or } \left(\frac{\partial P}{\partial V}\right)_V = \frac{R}{V} \qquad ...(7)$$

If pressure is constant, dP = 0; equation (6) becomes as follow :

$$\frac{R}{V}dT - \left(\frac{RT}{V^2} - \frac{2a}{V^3}\right)dV = 0$$

$$\text{or} \quad \left(\frac{\partial T}{\partial V}\right)_P = \left(\frac{RT}{V^2} - \frac{2a}{V^3}\right) \Big/ \frac{R}{V} \qquad ...(8)$$

If temperature is constant, dT = 0; equation (6) becomes as follows:

$$\left(\frac{\partial T}{\partial V}\right)_T = -\frac{1}{\left(\frac{RT}{V^2} - \frac{2a}{V^3}\right)} \qquad ...(9)$$

On multiplying equations (7), (8) and (9), we obtain

$$\left(\frac{\partial p}{\partial T}\right)_V \left(\frac{\partial T}{\partial V}\right)_P \left(\frac{\partial V}{\partial P}\right)_T = -\frac{R}{V} \times \frac{\left(\frac{RT}{V^2} - \frac{2a}{V^3}\right)}{\left(\frac{RT}{V^2} - \frac{2a}{V^3}\right)} \times \frac{1}{R} = -1$$

or $$\left(\frac{\partial p}{\partial T}\right)_V\left(\frac{\partial T}{\partial V}\right)_P\left(\frac{\partial V}{\partial P}\right)_T + 1 = 0$$

This is cyclic rule.

Example 63:

For an ideal gas (PV = nRT) show that (1/T) is an integrating factor for dw = P dV.

Solution:

$$dw = P\,dV \qquad ...(1)$$

$\because$ $$PV = nRT$$

$\therefore$ $$p\,dV + V\,dP = nR\,dT$$

and $$dV = \frac{nR}{P}dT - \frac{V}{P}dP \qquad ...(2)$$

On substituting the expression for dV in equation (1), we obtain

$$dw = nR\,dT - V\,dP$$
$$= nR\,dT - \frac{nRT}{P}dP$$

Now $$\frac{\partial}{\partial P}(nR)_T = 0$$

and $$\frac{\partial}{\partial T}\left(-\frac{nRT}{P}\right)_P = -\frac{nR}{P}$$

Hence, dw is inexact differential.

Suppose f = 1/T be an integrating factor. Then

$$f\,.\,dw = \frac{1}{T}nR\,dT - \frac{nR}{P}dP$$

$\therefore$ $$\frac{\partial}{\partial P}\left[\frac{nR}{T}\right]_T = 0$$

and $$\frac{\partial}{\partial T}\left[\frac{nR}{P}\right]_T = 0$$

Thus f. dw$\frac{nR}{T}dT - \frac{nR}{P}dP$ is an exact differential while 1/T is an integrating factor.

Example 64:

The opposite faces of a metal plate of 0.2 cm thickness an at a difference of temperature of 109°C and the area of the plate is 200 sq cm. Find the quantity of heat that will flow through the plate in one minute if K = 0.2 CGS units.

Solution:

Here,

$$K = 0.2$$

$$A = 200 \text{ sq cm}$$

$$d = 0.2 \text{ cm}$$

$$(\theta_1 - \theta_2) = 100°C$$

$$t = 60s$$

$$Q = \frac{KA(\theta_1 - \theta_2)t}{d}$$

$$= \frac{0.2 \times 200 \times 100 \times 60}{0.2}$$

$$= 12 \times 10^5 \text{ cal.}$$

Example 65:

A bar of length 30 cm and uniform area of cross-section 5 cm² consists of two halves AB of copper and BC of iron welded together at B. The end A is maintained at 200°C and the end C at 0°C. The sides of the bar are thermally insulated. Find the rate of flow of heat along the bar when the steady state is reached. Thermal conductivity of copper if 0.9 and thermal conductivity of iron is 0.12 CGS units.

Solution:

Suppose the temperature at the interface. B is θ after the steady state is reached.

$$\therefore \quad \frac{K_1A(200-\theta)}{d} = \frac{K_2A(\theta-0)}{d}$$

$$0.9(200 - \theta) = 0.12\,\theta$$

$$\theta = 176.5°C.$$

After the steady state is reached, the rate of flow of heat is the same in both the bars.

$$\therefore \quad Q = \frac{K_1A(\theta-0)}{d}$$

$$Q = \frac{0.9 \times 5 \times (200 - 176.5)}{15}$$

$$Q = 7.05 \text{ cal/s.}$$

Example 66:

Find the time in which a layer of ice 3 cm thick on the surface of a pond will increase its thickness by 1 mm when the temperature of the surrounding air is –20°C.

Thermal conductivity of ice = 0.005

Latent heat of ice = 80 cal/g

Density of ice at 0°C = 0.91 g/cm³.

Solution:

$$\int dt = \frac{\rho L}{K\theta} \int x dx$$

$$t = \frac{\rho L}{2K\theta}[x_2^2 - x_1^2]$$

Here,

$$\rho = 0.91 \text{ g/cm}^3$$

$$L = 80 \text{ cal/g}$$

$$K = 0.005$$

$$\theta = 20°C$$

$$x_1 = 3 \text{ cm}$$

$$x_2 = 3.1 \text{ cm}$$

$$t = \frac{0.91 \times 80}{2 \times 0.005 \times 20}[(3.1)^2 - (3)^2]$$

$$= 222.04 \text{ s}$$

$$= 3 \text{ min } 42 \text{ s approximately.}$$

Example 67:

Calculate the work down when 50 g of iron dissolves in hydrochloric acid at 25oC in :

(i) A closed vessel, and

(ii) An open beaker.

(iii) In which case would maximum amount of work be done ?

Solution:

$$\text{Amount of iron} = \frac{50 \text{ g}}{55.85 \text{ g mol}^{-1}} = \frac{50}{55.85} \text{ mol}$$

(i) In a closed vessel, $\Delta V = 0$. Hence,

$$dw = -PdV = 0$$

that is, no work is involved when iron dissolves in hydrochloric acid in a closed vessel.

(ii) The reaction between iron and hydrochloric acid is

$$Fe + 2HCl \rightarrow FeCl_2 + H_2(g)$$

Thus, the liberated hydrogen gas pushes the atmospheric air and thus involves the work of expansion. In this case, we will have

$$w = -P_{ext}\,\Delta V$$
$$= -P_{ext}\,V_{H2}$$

where V_{H2} is the volume of hydrogen released. According to the ideal gas law, we will have

$$w = -P_{ext}\,(n_{H2}RT/P_{ext})$$
$$= -n_{H2}RT$$
$$= -(50/55.85) \text{ mol } (8.314 \text{ J K}^{-1} \text{ mol}^{-1}) (298 \text{ K})$$
$$= -2.2 \text{ kJ.}$$

Example 68:

An ice box is built of wood 1.76 cm thick, lined inside with cork 3 cm thick. If the temperature of the inner surface of the cork is 0°C and that of the outer surface of wood is 12°C, what is the temperature of the interface ? The thermal conductivity of wood and cork are 0.0006 and 0.00012 CGS units respectively.

Solution:

Suppose the temperature of the interface is 9 after the steady state is reached

$$\therefore \quad \frac{K_1A(12-\theta)}{d_1} = \frac{K_2A(\theta-0)}{d_2}$$

Here $K_1 = 0.0006$,

$K_2 = 0.00012$

$$d_1 = 1.75 \text{ cm},$$

$$d_2 = 3 \text{ cm}$$

$$\frac{0.0006(12-\theta)}{1.75} = \frac{0.00012(\theta)}{3}$$

$$\theta = 10.74^{\circ}\text{C}.$$

Example 69:

Since CV = $(\partial q/\partial T)$ V by definition, one often writes without restriction dV = C_VdT. This is generally not true. Explain why.

Solution:

If we take U = f)V, T), then we will have

$$dU = \left(\frac{\partial U}{\partial V}\right)_T dV + \left(\frac{\partial U}{\partial T}\right)_V dT$$

$$= \left(\frac{\partial U}{\partial V}\right)_T . dV + C_V \, dT$$

Thus, if one writes dV = C_V dT without restriction then this statement is not complete. This is applicable for a system in which $(\partial U/\partial V)_T = 0$. An ideal gas is such an example. Alternatively, the system may be undergoing change at constant volume, for which, $(\partial U/\partial V)_T$ dV is zero and thus dV = C_V dT.

4

KINETIC MOLECULAR THEORY OF GASES

It is known that the gas laws are based o experiments quite independent of any theory of the nature of gas. However a simple gas theory that *a gas consists of a very large umber of molecules in rapid motion in a vacant space* can be very readily employed to derive and justify these laws mathematically. This theory is termed as the *Kinetic Molecular Theory of Gases.* This is named kinetic molecular theory because any model for explaining the behaviour of gases must involve both the molecular concept and motion of the particles. This theory was developed by Bernoulli, Clausius, Kronig, Maxwell, Kelvin, Boltzman and others. For the sake of generality, the theory assumes the properties of an ideal gas and is more or less based upon the following basic postulates:

POSTULATES OF THE KINETIC MOLECULAR THEORY OF GASES

Kinetic theory of gases is based upon the following assumptions:

(*i*) *All gases are made up of very large umber of minute particles called molecules.*

(*ii*) The molecules are small and separated from one another by large distances so that the actual volume occupied by the molecules is almost negligible as compared to the total volume of the gas.

(*iii*) The molecules are always moving in all directions in straight lines with very high velocities. They keep on colliding against each other (molecular collisions) and against the walls of the vessel at very small intervals of time.

(*iv*) The molecules are spherical and perfectly elastic, and there occurs o loss of kinetic energy due to mutual collisions or collisions with the walls of the container. The kinetic energy may be transferred from one molecule to another but it is not converted into any other form of energy such as heat.

(*v*) The pressure exerted by a gas on the walls of the containing vessel is due to the bombardment of the molecules of the gas on the walls of the containing vessel.

(*vi*) The average kinetic energy of the molecules of gas is directly proportional to the absolute temperature.

(*vii*) There are no attractive forces between molecules or between molecules and the walls of the vessel in which the gas is contained. The molecules move completely independent of one another.

(*viii*) The motion imparted to the molecules by gravity is negligible in comparison to the effect of the continued collisions between them.

(*ix*) The laws of classical mechanics (in particular the Newton's second law of option) are applicable to gaseous molecules.

Justification for the Postulates

The above postulates are sufficiently justified by the following experiences.

1. It is known that gases can be compressed by the application of pressure and can also expand indefinitely filling the whole vessel no matter how large it may be. This reveals that gases are made of very large umber of minute particles separated from each other by large distances. Further these particles occupy very small volumes as compared to the total volume of the gas. In a gas like oxygen, hydrogen or nitrogen, at N.T.P. the volume occupied by the molecules themselves is only 0.014 percent of the total volume. The remaining 99.986 per cent of the volume is merely an empty space.

   [Postulates (*i*) and (*ii*)]

2. Dust particles i air are found to be in constant irregular zig-zag motion termed *Brownian movement*. This is an evidence for the very fast straight line motion of the molecules.[Postulate (*iii*)]

3. Rapid motion of a large number of molecules must involve frequent collisions between then and these collisions must be elastic as otherwise there will occur constant loss of kinetic

energy and the molecules must come to stand still which does not happen. [Postulates (*iii*) and (*iv*)]

4. If a gas is contained in a closed vessel, "it is expected that moving molecules will keep on hitting the walls of the vessel continuously thus exerting pressure." [Postulate (*v*)]
5. Molecules are very close in a compressed gas and intermolecular attraction should be high but this attraction must be negligible because even a highly compressed gas expands freely and fills the whole available space.

   [Postulate (*vii*)]
6. Gases intermix freely with other gases in all proportions. Even a heavier gas rises up to intermix with a lighter gas. This indicates that the effect of gravity is negligible i comparison to the effect of the continued collisions between them. [Postulate (*viii*)]
7. On heating, the temperature of gas rises. Heat energy supplied increases the speed of the molecules and their average kinetic energy increases thereby causing the rise of temperature. [Postulate (*vi*)]

Derivation of the Kinetic Equation for Gases

BY taking the postulates of kinetic theory into consideration, it is possible by applying the laws of classical mechanics to derive a mathematical expression which can explain the various gas laws. This expression is known as the *Kinetic Gas Equation.*

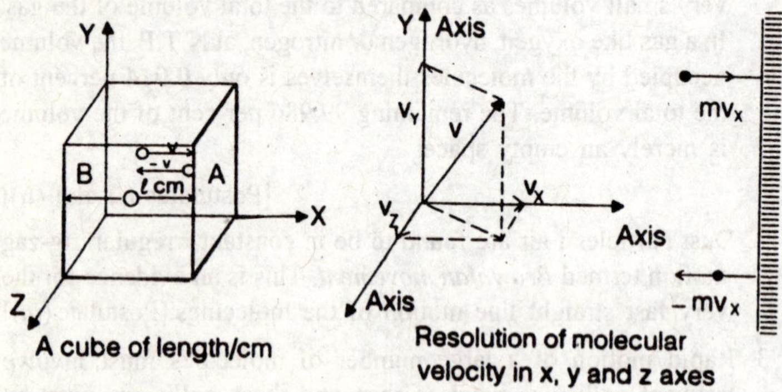

Fig. 4.1

Suppose a definite mass of a gas is contained in a cobical vessel (Fig. 4.1) and suppose that

the length of each side of the cube = l cm

the total number of molecules = N

the mass of one molecule = m

The selection of the cubical shape of the container is purely a matter of convenience and involves no loss of generality because the pressure of a gas is independent of the shape of the container.

It is regarded that the container are smooth and get oriented perpendicular to the x, y and z-axes of a Cartesian coordinate system. The motion of molecules in the container, at any instant, is totally random.

Although, the molecules are moving in every possible direction the average velocity v of a molecule at any moment can be resolved into three directions X, Y, Z at right angles to one another (Fig. 4.1).

Suppose a single molecule is moving with a velocity v.

Suppose v_x, v_y and v_z be the components of velocity v in the x, y ad z directions parallel to the three edges of the cube.

These three components get related to the resultant velocity v by the expression give below:

$$v = v_x^2 + v_y^2 + v_z^2 \quad ...(1)$$

Suppose the motion of this molecule occurs between the two opposite faces A and B parallel to the x-axis. It strikes the face A at regular intervals.

The velocity of the molecule before it strikes the face A is v_x. But the molecule is perfectly elastic. Therefore it will rebound with the same velocity with the sign changed, *i.e.*, $-v_x$.

$$= 2mv_x \text{ gm. cm/sec} \quad ...(2)$$

In order to strike the same face (A) against, the molecule will have to move to opposite face 'B' and come back. It means that the particle has to travel a distance 2l cm for each successive collision on the same face.

Thus, by traversing 2l cm., one impact is produced. By traversing l cm., 1/2l impacts get produced.

$\therefore$ By traversing v_x cm, $\frac{v_x}{2l}$ impacts get produced per second.

(v_x is velocity of the molecule that is the distance travelled i one second.)

Hence the number of impacts made by one molecule in one second

$$= \frac{v_x}{2l}$$

$\therefore$ Total change ion momentum per second (due to $\frac{v_x}{2l}$ impact) on face A

$$= 2mv_x \times \frac{v_x}{2l}$$

$$= \frac{mv_x^2}{l} \text{ gm cm/sec}^2$$

Total change in momentum per second because of the impacts of the molecule on two opposite faces A and B, along x-axis is therefore, double, *i.e.*, equal to

$$\frac{mv_x^2}{l} + \frac{mv_x^2}{l} = \frac{2mv_x^2}{l} \qquad ...(3)$$

Also the total change of momentum per second due to impact of a single molecule on the two opposite faces along y-axis and two opposite faces along z-axis will be give as follows:

$$\frac{2mv_y^2}{l}, \frac{2mv_z^2}{l} \text{ respectively.}$$

Therefore, the total change in momentum per second o all the six faces of the cube, is equal to

$$= \frac{2mv_x^2}{l} + \frac{2mv_y^2}{l} + \frac{2mv_z^2}{l}$$

or $$= \frac{2m}{l}(v_x^2 + v_y^2 + v_z^2)$$

or $$= \frac{2mv^2}{l} \text{ gm cm/sec}^2 \qquad ...(4)$$

$$(\because \quad v_x^2 + v_y^2 + v_z^2 = v^2)$$

Thus total change of momentum, per second due to one molecule

$$= \frac{2mv^2}{l} \text{ gm cm/sec}^2$$

According to Newton's second law of motion, the rate of momentum is force *i.e.*,

$$\therefore \qquad \text{Force} = \frac{2mv^2}{l} \text{ dynes}$$

The pressure 'P' (force per unit area) is the total force divided by area of six faces of the cube ($6l^2$), *i.e.*,

$$P = \frac{\text{Force}}{6l^2} = \frac{2mv^2}{l} \times \frac{1}{6l^2} = \frac{1}{3}\frac{mv^2}{l^3}$$

or
$$P = \frac{1}{3}\frac{mv^2}{l^3} \qquad \text{...(5)}$$

But l^3 is the volume of the cube 'V'

$$P = \frac{1\, mv^2}{3\, V}$$

Hence Pressure exerted by one molecule is

$$P = \frac{1}{3V} mv^2 \qquad \text{...(6)}$$

The total pressure exerted by N different molecules of same mass ad having different velocities v_1, v_2, v_3 ... v_N will be as follows:

$$P = P_1 + P_2 + P_3 + ... + P_N$$

$$= \frac{1}{3V} mv_1^2 + \frac{1}{3V} mv_2^2 + \frac{1}{3V} mv_3^2 + ... + \frac{1}{3V} mv_N^2$$

[from Equation ... (6)]

$$P = \frac{1}{3V} m\,(v_1^2 + v_2^2 + v_3^2 + ... + v_N^2) \qquad \text{...(7)}$$

If we multiply and divide the right had side of equation (7) by , we get

$$P = \frac{mN}{3V} \frac{(v_1^2 + v_2^2 + v_3^2 + ...\, v_N^2)}{N}$$

$$= \frac{mN}{3V} \times (\text{mean square velocity})$$

or $$P = \frac{mN}{3V} \times (\text{root mean square velocity})^2$$

$$= \frac{mNc^2}{3V}$$

where N refers to the total number of molecules and $\sqrt{\bar{c}^2}$, the root mean square velocity give by

$$\sqrt{\bar{c}^2} = \sqrt{\frac{v_1^2 + v_2^2 + v_3^2 + ... v_N^2}{N}}$$

$$PV = \frac{1}{3} mN\bar{c}^2 \quad ...(8)$$

This is known as the Kinetic Gas Equation

It may be expressed as $PV = \frac{1}{3}M\bar{c}^2$ where M is the total mass of the gas.

or $$P = \frac{1}{3}\frac{m}{V}\bar{c}^2 = \frac{1}{3}D\bar{c}^2$$

where D is density of the gas.

Kinetic Energy and Temperature

Translational kinetic energy of a body of mass m with velocity $\sqrt{\bar{c}^2}$ is give by as follows:

$$K.E. = \frac{1}{2}m\bar{c}^2$$

As the velocity of a molecule increases with the rise in temperature, it becomes possible to define mean kinetic energy in terms of temperature of the gas.

Suppose one mole of an ideal gas is under consideration. Then the total number of molecules is N (Avogadro's number).

∴ Kinetic energy per mol of the gas for N particles is given by

$$K.E. = \frac{1}{2}mN\bar{c}^2 \quad ...(9)$$

The kinetic gas equation (8) for N molecules of a gas is

$$PV = \frac{1}{2}mN\bar{c}^2$$

or $$PV = \frac{2}{3} \times \frac{1}{2} mN\bar{c}^2$$

or $$= \frac{2}{3} K.E. \qquad (\because \frac{1}{2} mNc^2 \text{ is the kinetic energy})$$

On comparing this equation with ideal gas equation PV = RT for 1 mole of the gas, we obtain

$$RT = \frac{2}{3} K.E.$$

or $$K.E. = \frac{3}{2} RT$$

or $$K.E. \propto T \qquad ...(10)$$

Hence the kinetic energy of one mole of any gas is directly proportional to its absolute temperature.

This is also the Maxwell's generalisation, which states that "*Kinetic energy of translation of an ideal gas is independent of the nature of the gas, its pressure and depends only upon the temperature.*"

But $$K.E. \propto \bar{c}^2$$

$$\therefore \quad \bar{c}^2 \propto T$$

or $$\bar{c}^2 \propto \sqrt{T} \qquad ...(11)$$

Hence molecular velocity of any gas is proportional to the square root of the absolute temperature. The molecular motion is, therefore, often known as *thermal motion of the molecules.*

At absolute zero, the thermal motion ceases completely and kinetic energy of the gas would become also zero.

Deviation of Gas Laws from the Kinetic Equation

(a) Boyle's Law: *This law states, "Temperature remaining constant, the volume of a give mass of a gas is inversely proportional to the pressure."* mathematically it can be expressed as

$$V \propto \frac{1}{P}$$

$$PV = \text{constant}$$

This law can be derived from the kinetic equation

$$PV = \frac{1}{3} mN\bar{c}^2$$

This can be written as

$$PV = \frac{2}{3} \times \frac{1}{2} mN\bar{c}^2 \quad ...(i)$$

But $\frac{1}{2} mN\bar{c}^2$ = average kinetic energy of gas which is proportional to the absolute temperature.

i.e. $$\frac{1}{2} mN\bar{c}^2 \propto T$$

or $$\frac{1}{2} mN\bar{c}^2 = KT$$

Substituting in (i) above

$$PV = \frac{2}{3} . KT$$

In this equation $\frac{2}{3}$ KT would be a constant for a give mass of gas at a constant temperature.

$\therefore$ $$PV = \text{constant}$$

or $$V \propto \frac{1}{P}.$$ This is Boyle's law.

(b) Charle's Law: *According to this law, "The volume of a give mass of a gas is directly proportional to the absolute temperature at a constant pressure."*

We have already derived

$$PV = \frac{2}{3} kT$$

or $$V = \frac{2}{3} \times \frac{k}{P} . T$$

or $$\frac{V}{T} = \frac{2k}{3P}$$ Thus for a give mass of a gas

if pressure P, remains constant, the expression $\frac{2k}{3P}$ is constant.

$\therefore$ At constant pressure

$$\frac{V}{T} = K$$

or $$V = K.T.$$

Hence $V \propto T.$ This is Charle's law.

(c) Avogadro's Hypothesis: *This law states, "Equal volumes of all gases under similar conditions of temperature ad pressure contain equal number of molecules."*

For any two gases, the kinetic equation can be put as follows:

(i) $$P_1V_1 = \frac{1}{3} m_1N_1\bar{c}_1^2$$

(ii) $$P_2V_2 = \frac{1}{3} m_1N_1\bar{c}_1^2$$

On rewriting:

(i) $$P_1V_1 = \frac{2}{3} \times \frac{1}{2} m_1N_1\bar{c}_1^2$$

(ii) $$P_2V_2 = \frac{2}{3} \times \frac{1}{2} m_1N_1\bar{c}_2^2$$

Since $P_1 = P_2$ and $V_1 = V_2$

$$\therefore \quad \frac{1}{2} m_1N_1\bar{c}_1^2 = \frac{1}{2} m_1N_1\bar{c}_2^2 \quad \text{...(a)}$$

Now if the temperature is also constant, the average kinetic energy per molecule must be the same.

or $$\frac{1}{2} m_1\bar{c}_1^2 = \frac{1}{2} m_2\bar{c}_2^2 \quad \text{...(b)}$$

On dividing (a) by (b), we have

$$N_1 = N_2$$

This is Avogadro's Law

(d) The Ideal Gas Equation: On Combining Boyle's law, Charles law ad Avogadro's law we find that the volume of a gas depends on the pressure, temperature and number of moles, as follows:

$V \propto 1/P$ (at constant T and n) (Boyle's law)

$V \propto T$(at constant P and n) (Charles' law)

$V \propto n$(at constant T and P) (Avogadro's law)

Thus, V should be proportional to the product of these three terms, *i.e.*,

$$V \propto nT/P$$

or $$V = R(nT/P)$$

or $$PY = RT \quad ...(i)$$

where R, the proportionality constant, is called the gas constant. Eq. (i) is called the ideal gas equation.

The Gas Constant: As one mole of a ideal gas occupies a volume of 22.414 dm^3 at N.T.P. (0°C ad 1 atmp pressure), we find from Eq. (i) the gas constant

$$R = \frac{PV}{nT} = \frac{(1\,\text{atm})(22.414\,\text{dm}^3)}{(1\,\text{mol})(273.15\,\text{K})}$$

$$= 0.08206\ \text{dm}^3\ \text{atm}\ \text{K}^{-1}\ \text{mol}^{-1}$$

In SI units, 1 atom = $1.01325 \times 10^5\ \text{N m}^{-2}$, so that

$$R = \frac{PV}{nT}$$

$$= \frac{(1.03125 \times 10^5\,\text{N m}^{-2})(22.414 \times 10^{-3}\,\text{m}^3)}{(1\,\text{mol})(273.15\,\text{K})}$$

$$= 8.314\ \text{N m K}^{-1}\ \text{mol}^{-1}$$

$$= 8.314\ \text{J K}^{-1}\ \text{mol}^{-1} \qquad (\because\ J = Nm)$$

Note: 1 dm^3 = $1 \times 10^{-3}\ m^3$ = 1000 cm^3 = 1 litre.

(e) Dalton's Law of partial pressure: This law states, "*Total pressure P of a number of gases i a enclosed space is equal to the sum of the individual (partial) pressures of each gas, provided the gases do not react with one another.*" Thus

$$P = p_1 = p_2 + p_3 + p_4 + ... + p_N$$

Gas pressure is the force per unit area exerted by the molecules of the gas against the walls of the container.

Suppose N_1 molecules, each of mass m_1 of a gas A, are contained in a vessel of volume V.

The according to the kinetic theory equation the pressure is given by

$$p_a = \frac{m_1 N_1 \bar{c}_1^2}{3V} \quad \text{...(i)}$$

where $\bar{c}_1$ is the root mean square velocity of the molecules of the gas A.

Now, suppose N_2 molecules, each of mass m_2 of another gas B, are contained i the same vessel at the same temperature and there is no other gas present at that time. The pressure of the gas will b given by

$$p_b = \frac{m_2 N_2 \bar{c}_2^2}{3V} \quad \text{...(ii)}$$

where $\bar{c}_2$ is root mean square velocity of the molecules of the gas B.

If both the gases are present in the same vessel, the total pressure P will be give by

$$P = \frac{m_1 N_1 \bar{c}_1^2}{3V} + \frac{m_2 N_2 \bar{c}_2^2}{3V} = p_a + p_b$$

Similarly if three, four or ore gases were present, the total pressure P will be given by

$$P = p_a + p_b + p_c + p_d \ldots$$

This is Dalton's law of partial pressures.

(f) Gaham's Law of Gaseous Diffusion: The law may be stated "*Under similar conditions of temperature and pressure, the rates of diffusion of gases are inversely proportional to the square roots of their densities.*"

Mathematically,

$$\frac{r_1}{r_2} = \sqrt{\frac{D_2}{D_1}}$$

According to the kinetic equation

$$PV = \frac{1}{3} m N \bar{c}^2$$

or

$$\bar{c} = \sqrt{\frac{3PV}{mN}}$$

$$= \sqrt{\frac{3PV}{M}} \quad \text{[mN = molecular weight of the gas.]}$$

$$= \sqrt{\frac{3P}{D}} \quad [\because \ D = \frac{M}{V}]$$

$$\bar{c} = \sqrt{\frac{1}{D}} \text{ if P is kept constant.}$$

$\bar{c}$ is the mean velocity of the molecules and rate of diffusion of a gas will be dependent upon it at a particular temperature.

$$r \propto \bar{c}$$

$$r \propto \sqrt{\frac{1}{D}} \text{ if P is kept constant.}$$

$\bar{c}$ is the mean velocity of the molecules and rate of diffusion of a gas will be dependent upon it at a particular temperature.

$$r \propto \bar{c}$$

$$r \propto \sqrt{\frac{1}{D}}$$

This is Graham's Law.

Calculation of Molecular Velocity

In calculating various properties of the gases, one makes use of the following three velocities.

(i) Root mean Square velocity (RMSV): *It may be defined as the square root of the mean of the velocities of the molecules.* If v_1 v_2 ... etc. are the velocities possessed by different molecules, then

$$\sqrt{\bar{c}^2} = \sqrt{\frac{v_1^2 + v_2^2 + \ldots + v_N^2}{N}}$$

The expression to calculate this velocity can be obtained from the kinetic gas equation, according to which

$$PV = \frac{1}{3} mN\bar{c}^2$$

For 1 mole of the gas

$$PV = RT \text{ and } N = NA \text{ thus}$$

$$\sqrt{\bar{c}^2} = \sqrt{\frac{3PV}{mNa}} = \sqrt{\frac{3RT}{mNa}} = \sqrt{\frac{3RT}{M}} = \sqrt{\frac{3kT}{m}} = \sqrt{\frac{3P}{d}}$$

where K is Boltzmann constant k = R/Na

(2) Average Velocity (c): *It is defined as the average of the various velocities possessed by the molecules, i.e., as*

$$\overline{c} = \frac{v_1 + v_2 + ... + v_N}{N}$$

It can be calculated by using the expression

$$\overline{c} = \sqrt{\frac{8RT}{\pi M}}$$

$$\overline{c} = \sqrt{\frac{8RT}{\pi M}} = \sqrt{\frac{8kT}{\pi m}}$$

(3) Most Probably Velocity (MPV): It may be defined *as the velocity corresponding to the maximum fraction of the molecules.* It is given by the following expression

$$MPV = \sqrt{\frac{2RT}{M}} = \sqrt{\frac{2PV}{M}} = \sqrt{\frac{2P}{d}}$$

The relations between these velocities can be easily worked out

$$\frac{MPV}{\overline{c}^2} = \frac{\sqrt{2RT/M}}{\sqrt{8RT/\pi M}} = \sqrt{\frac{22}{28}} = \frac{1}{1.128}$$

$$\frac{MPV}{RMSV} = \frac{\sqrt{2RT/M}}{\sqrt{3RT/M}} = \sqrt{\frac{2}{3}} = \frac{1}{1.224}$$

Thus MPV : $\overline{c}$: RMSV : : 1 : 1 . 128 : 1.224.

$$\textbf{Average} = \left(\frac{1.128}{1.224}\right) \textbf{MPV} = \textbf{0.9213 MPV}$$

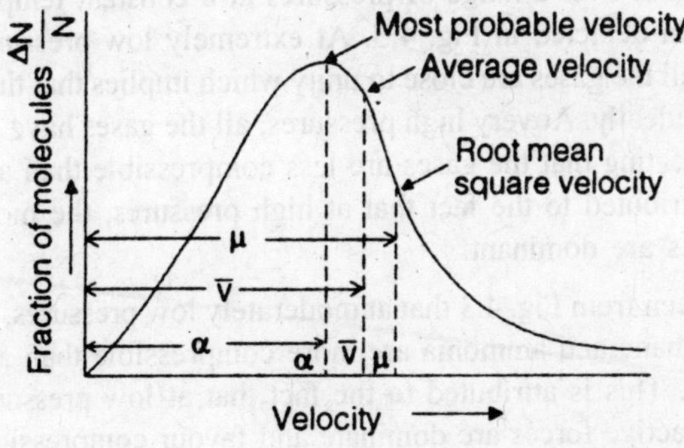

Fig. 4.2 : Relative values of most probable velocity, Average ($\overline{c}$) velocity $\sqrt{c^2}$ and Root mean square velocity.

i.e. RMSV > c > MPV. This is also shown in Fig. (4.2). The relation between average and RMSV will be

$$\text{Average} = \left(\frac{1.128}{1.224}\right) \text{MPV} = 0.9213 \text{ MPV}$$

COMPRESSIBILITY FACTOR

It is known that the equation of state, PV = nRT, derived from the postulates of the kinetic theory, *is valid for an ideal gas only*. Real gases tend to obey this equation only approximately ad that too under conditions of low pressure and high temperature.

The higher the pressure and the lower the temperature, the greater would be the deviations from the ideal behaviour. In general, the most easily liquefiable and highly soluble gases exhibit larger deviations. Hence gases like carbon dioxide, sulphur dioxide and ammonia exhibit much larger deviations than hydrogen, oxygen, nitrogen, etc.

It is best to represent the deviations from ideal behaviour i terms of the compressibility factor, Z, which is defined as

$$Z = \frac{PV}{(PV)_{ideal}} = \frac{PV}{nRT} \qquad ...(1)$$

For an ideal gas, Z = 1 under all conditions of temperature and pressure. The deviation of Z from unity is, therefore a measure of the imperfection of the gas under consideration.

The graphs plotted for the compressibility factors determined for a number of gases over a range of pressures at a constant temperature (0°C) have been depicted in Fig. 4.3. At extremely low pressures, the value of Z for all the gases are close to unity which implies that the gases behave almost ideally. At very high pressures, all the gases have Z more than unity indicating that the gases are less compressible than an ideal gas. This is attributed to the fact that at high pressures, the molecular repulsive forces are dominant.

It can be seen from Fig. 4.3 that at moderately low pressures, carbon monoxide, methane and ammonia are more compressible than an ideal gas, *i.e.* Z < 1. This is attributed to the fact that at low pressures, the long range attractive forces are dominate and favour compression. The compressibility factor Z goes on decreasing with increase in pressure, passes through a minimum at a certain stage and then starts increasing with increase in pressure. The gases would now becomes less compressible than ideal gas, *i.e.*, Z > 1. It can be seen from the figure that although carbon monoxide and methane exhibit marked devications from ideal behaviour only at high pressures, yet ammonia exhibits large deviation even at low pressure.

It can be seen that hydrogen and helium at 0°C are less compressible *at all pressures* than the ideal gas, *i.e.*, Z > 1. However, if the temperature is sufficiently low (*e.g.*, below – 165°C for hydrogen and below – 240°C for helium) these gases would also give the same type of Z – P plots as are shown by ammonia, carbon monoxide and methane at 0°C.

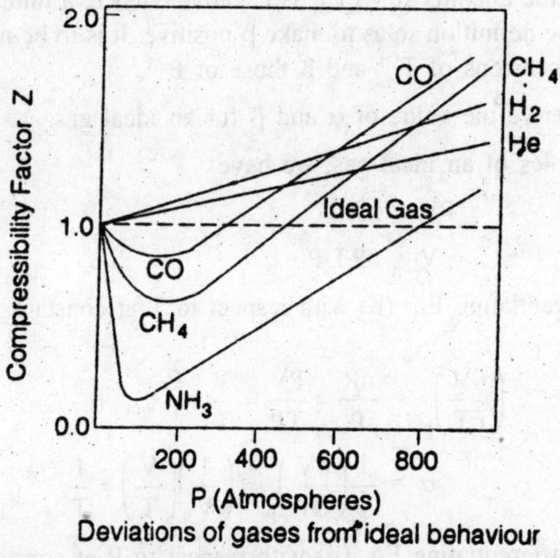

Fig. 4.3 : Deviations of gases from ideal behaviour.

On the other hand, if the temperature becomes sufficiently high, the Z-P plots of ammonia, carbon monoxide and methane will be almost similar to those of hydrogen and helium at 0°C *i.e.*, the value of Z will increase continuously with increasing pressure.

EXPANSIVITY AND COMPRESSIBILITY

When fluids (gases and liquids) are heated, they expand but the expansion of gases is much more than that of liquids. Similarly, they can be compressed. The variation of volume V with temperature T, keeping pressure P constant is termed as coefficient of thermal expansion or the coefficient of isobaric expansion or simple, expansivity, α of the fluid. Thus,

$$\alpha = \frac{1}{V}\left(\frac{\partial V}{\partial T}\right)_P \quad ...(1)$$

Similarly, the variation of V with P, keeping T constant, is termed *as the coefficient of isothermal compressibility* or, simply, compressibility, β of the fluid, thus,

$$\alpha = -\frac{1}{V}\left(\frac{\partial V}{\partial T}\right)_T \quad ...(2)$$

Generally, as the volume of a gas gets decreased as pressure increases it means that the quantity $(\partial V/\partial P)_T$ is negative. Hence, a minus sign is included in the definition so as to make β positive. It is to be noted that α has the dimensions of T^{-1} and β those of P^{-1}.

Let us derive the value of α and β for an ideal gas.

For n moles of an ideal gas, we have

$$PV = nRT \quad ...(i)$$

or

$$V = nRT/P \quad ...(ii)$$

On differentiating Eq. (ii) with respect to T at constant P, we get,

$$\left(\frac{\partial V}{\partial T}\right)_P = \frac{nR}{P} = \frac{PV}{TP} = \frac{V}{T}$$

$$\therefore \quad \alpha = \frac{1}{V}\left(\frac{\partial V}{\partial T}\right)_P = \left(\frac{1}{V}\right)\left(\frac{V}{T}\right) = \frac{1}{T}$$

Again, differentiating Eq. (ii) with respect to P at constant T. We get,

$$\left(\frac{\partial V}{\partial P}\right)_T = \frac{nRT}{P^2} = \left(\frac{nRT}{P}\right)\left(-\frac{1}{P}\right) = -\frac{V}{P}$$

$$\therefore \quad \beta = -\frac{1}{V}\left(\frac{\partial V}{\partial P}\right)_T = \left(\frac{1}{V}\right)\left(-\frac{V}{P}\right) = \frac{1}{P}.$$

DEVIATION FROM IDEAL GAS BEHAVIOUR

If we study the actual behaviour of the gases, it can be see that no gas *is ideal or perfect* in this sense although some gases tend to approach perfection as the temperature rises above their boiling points.

Since we know that no gas is ideal or perfect, the equation PV = nRT is obeyed as an approximation by real gas at low pressure while at high pressure and low temperature there occurs much deviation from this ideal behaviour.

Regnault and Amagat attempted to study the effect of change of P on PV of real gases like H_2, He, N_2O_2 and CO_2 under different conditions of temperatures and demonstrated that they did not obey Boyle's law, more particularly at high pressure and low temperatures. Their results have been depicted in Fig. 4.4.

From the diagram the following conclusions can be drawn:

(i) *At low Pressure:* Left hand portions of the curves show the devications at low pressures. With the rise of pressure, PV increases for hydrogen and helium whereas it decreases for 23 and CO2. Same is the case with other gases. *Thus at low pressure PV for all gases excepting H_2 and He, is having a lower value than expected of an ideal gas.*

(ii) *At high pressure:* Right-had portions of the curves show the deviations at high pressures. For H_2 and He, PV gets increased regularly with the rise of pressure but for N_2 and CO_2 it first decreases, passes through a minimum where it remains constant for a while and the rises. At very high pressure PV for all gases is having a higher value than that expected of an ideal gas.

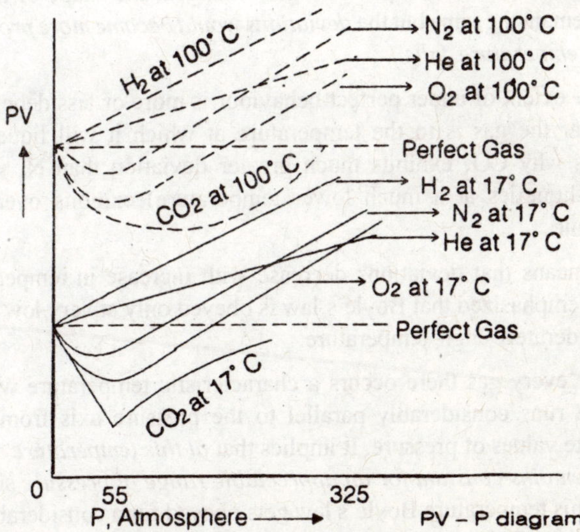

Fig. 4.4 : PV-P diagram.

Thus at high pressures PV for all gases including H_2 and He is having higher value than expected for an ideal gas.

The above results can be interpreted as follows:

(a) At moderate pressures, there occur a negative or under perfect deviation. Hydrogen and helium do not appear to show under perfect behaviour at ordinary temperature. However, at a much lower temperature hydrogen and helium also give a curve of thc same type as the other gases.

(b) All the curves approach the ideal value as the pressure approaches zero, *i.e.*, at low pressure all gases behave as ideal gases.

(c) At very high pressures, there occurs a positive or over perfect devication.

(d) At very high pressures gases do not eve approximately obey gas laws.

(iii) *Effect of temperature:* In Fig. 4.1, there are two sets of curves, one at 170186 and the other at 100°C. It is evident from these curves that the deviations are greater at 17°C while in case of CO_2 a sharp break occurs as the liquefaction point gets reached.

In general if the temperature gets changed, the shape of the curve would remain the same but the *deviations would become more pronounced as the temperature falls.*

The extent of under perfect behaviour is more or less dependent on how near the gas is to the temperature at which it will liquefy. This explains why CO_2 exhibits much greater deviation than N_2 while H_2 (which liquefies at a much lower temperature) exhibits over perfect behaviour.

It means that deviations decrease with increase in temperature. It may be emphasized that Boyle's law is obeyed only at very low pressure and moderately high temperature.

For every gas there occurs a characteristic temperature when PV-P curve runs considerably parallel to the pressure axis from zero to moderate values of pressure. It implies that *at this temperature the value of PV remains constant for an appreciable range of pressure* signifying that at this temperature Boyle's law gets obeyed for a considerable range of pressure. This temperature is termed as the Boyle's temperature.

Explanation of Deviation

So as to explain deviations from ideal behaviour, it becomes essential to modify the kinetic theory of gases. The following two postulates of the kinetic theory do not hold good under all conditions. We will make an attempt to examine them more critically.

Postulate No. 1: The volume occupied by the gas molecules themselves is negligibly small as compared to the total volume occupied by the gas. One can justify this postulate only under ordinary conditions of temperature and pressure.

By calculations it can be see that in some of the common gases, the volume occupied by the molecules themselves, under ordinary conditions, is early 0.014 percent of the total volume of the gas. This is a negligible fraction. But, if the pressure becomes too high (say, 100 atmosphere or more), the total volume of the gas will decrease appreciably while the volume of the molecules will remain almost the same because the molecules are incompressible. *Therefore under conditions of high pressure, the volume occupied by the gas molecules cannot be neglected in comparison with the total volume of the gas.*

The same thing is expected to happen when the temperature is decreased to a large extent. The total volume of the gas gets decreased considerably, no doubt, but the volume occupied by the molecules themselves remains practically the same.

In this case, too, *the volume occupied by the molecules would be no longer be negligible. Thus the postulate No. 1. is not valid at high pressures ad low temperatures.*

Postulate No. 2 : The forces of attraction between gas molecules are negligible. This assumption is expected to be valid at low pressures or at high temperatures because under these conditions the molecules li far apart from once another.

But at high pressures or at low temperatures, the volume becomes small and molecules lie closer to one another. The intermolecular forces of attraction, therefore, become appreciable and cannot be ignored.

Hence, *the postulate No. 2, is not valid under conditions of high pressure and low temperature.*

It becomes necessary, therefore, to apply suitable corrections to the ideal gas equation so as *to make it applicable to real gases.*

EQUATIONS OF STATE FOR REAL GASES

A number of equations of state have been given to describe the P-V-T relationship in real gases. The earliest ad the best known equation is the one which is given by van der Waals.

The van der Waals Equation of State

In 1873, van der Waals gave his famous equation of state for a non-ideal (*i.e.*, imperfect gas). He modified the ideal gas equation by postulating that the gas molecules were not mass points but behaved like rigid spheres having a certain diameter and that there existed *intermolecular forces of attraction between them*. The two correction terms introduced by van der Waals have been described below.

1. *Correction due to volume of gas molecules:* It is known that the derivation of the ideal gas equation PV = nRT is based on the assumption that the gas molecules are mass points, *i.e.*, they do not have any finite volume. van der Waals discarded this assumption and suggested that a correction term *nb* should be subtracted from the total volume V so as to get the *ideal volume which is compressible.*

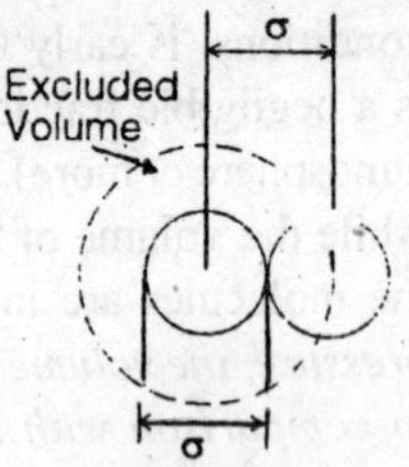

Excluded Volume

σ

σ

Illustration of excluded volume

Fig. 4.5

In order to understand the meaning of the correction term b, consider two gas molecules as unpentrable and incompressible spheres, each of which is having a diameter σ as shown in Fig. 4.5. It can be seen that the centres of the two spheres cannot approach each other more closely than the distance σ.

For this pair of molecules, therefore, a sphere of radius σ and therefore of volume $\frac{4}{3}\pi\sigma^3$ constitutes what is termed as the excluded volume. The excluded volume per molecule would be therefore half the above, volume, viz., $\frac{2}{3}\pi\sigma^3$. The actual volume of one gas molecule of radius r is $\frac{4}{3}\pi\sigma^3 = \frac{4}{3}\pi(\sigma/2)^3 = \frac{1}{6}\pi\sigma^3$.

∴ Excluded volume per molecule = $\frac{2}{3}\pi\sigma^3 = 4 \times \frac{1}{6}\pi\sigma^3$ = 4 times the actual volume oft he gas molecule.

The excluded volume would be *thus 4 times the actual volume of the molecule*. Evident, the excluded volume per mol of the gas would be $N_A \times 4 \times \frac{4}{3}\pi r^3 = b$ where N_A refers to the Avogadro's number. The compressible volume per mole of the gas would thus be V – b. If volume V of the gas is having n moles, the excluded volume would be nb. Thus, the ideal volume which is compressible is V-nb, the volume b per mole is also termed as co-volume.

2. *Correction due to intermolecular forces of attraction:* In the derivation of the ideal gas equation, an assumption has been made that there are no intermolecular forces of attraction, we will consider a molecule which is lying somewhere in the midst of the vessel, as shown at the point A (Fig. 4.6). Thus molecule is being attracted uniformly on all sides by the neighbouring molecules. These forces are able to neutralize one another ad there occurs no resultant attractive force on the molecule. However as the molecule approaches the wall of the vessel, as shown at B, it would experience attractive forces form the bulk of the molecules behind it.

Hence, it means that it will strike the wall with a lower velocity and will exert a lower pressure than it would have done if there was no force of attraction. It is, therefore, necessary to add a certain quantity to the pressure of the gas i order to obtain the *ideal pressure.* The correct pressure, therefore, world be P + p.

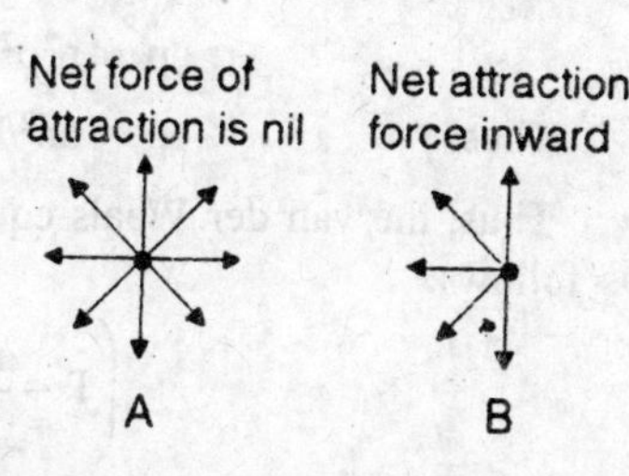

Fig. 4.6

Calculation of Correction Factor p

The force of attraction exerted on a single molecule which is about to strike the wall evidently is dependent upon the *number of molecules per unit volume in the bulk of the gas, i.e.*, directly upon the *density* of the gas. Further, the number of molecules striking the wall at any given instant is also dependent directly upon the *density* of the as. Thus, the total inward attractive pull on the molecules which gives a measure of the correction factor, p, is proportional to the square of the density (ρ) of the gas, *i.e.*,

$$p \propto \rho^2 \qquad ...(2)$$

But density is inversely proportional to the volume and if V refers to the volume occupied by one mole of a as, the value of p for one mole of a gas will be inversely proportional to the square of volume.

Hence $p \propto 1/V^2$

or $p = a/V^2$ ...(3)

where a refers to a *constant* depending upon the nature of the gas.

The kinetic gas equation for *one mole* of a real gas, therefore, takes the following form

$$(P + a/V_m^2)(V_m - b) = RT \quad ...(4)$$

where V_m refers to the *molar volume* of the gas.

This is termed as van der Waals equation. The constant a and b are termed as van der Waals constants. These are characteristic of each gas.

Eq. (4) has been found to be valid for one mole of a gas only. If there are moles of a gas occupying volume V, then, the excluded volume will be given by b and the compressible volume, therefore, will be V – nb. The pressure correction factor p for n moles, in the light of Eq. 2, will be proportional to $n^2\rho^2$, *i.e.*

$$p \propto n^2\rho^2 \ \mu \ n^2 \times 1/V^2$$

Thus, $p = an^2/V^2$ ...(5)

Thus, the van der Waals equation for n moles of a as may be put as follows:

$$\left(P + \frac{n^2a}{V^2}\right)(V - nb) = mRT \quad ...(6)$$

Table 4.1 : Volume of Carbon Dioxide at 47°C under Different Pressures.

Pressure (atm)	*Volume (dm^3) of the gas*		
	Observed experimentally	*Calculated from van der Waals equation*	*Calculated from ideal gas equation*
1	27.20	26.20	26.30
10	2.52	2.53	2.63
40	0.54	0.55	0.66
100	0.098	0.10	0.20

Eq. 6 has been found to be more accurate than the ideal as equation PV = nRT for expressing the PVT behaviour of real gases. Thus, if one mol of carbondioxide is taken at 47°C and compress it to different pressures, the volume, as observed by experiment, is found to be closer to that calculated from van der Waals equation that to that calculated from the ideal gas equation. The departure from ideal gas equation would become more and more wide as the pressure continues to rise (Table 4.1).

Units for van der Waals Constants

The unit for the van der Waals constants a and b would depend upon the units in which *P* and *V* are expressed. It follows from Eq. 5 that the constant a is expressed by pV^2/n^2, *i.e.* pressure × $(volume)^2/mol^2$. If pressure is expressed in atmospheres and volume in dm^3 the value of a will be in atm dm^6 mol^{-2}. As regards b, it is incompressible volume per mole, of a gas. Hence it must have the same units as volume per mole, e.g, dm^3 mol^{-1}.

The van der Waals constants for some gases have been included in Table 4.2.

Table 4.2 : van der Walls constants for some common gas.

Gas	*a (dm^6 atm. mol^{-2})*	*b (dm^3 mol^{-1})*
Ammonia	4.17	0.0371
Argon	1.35	0.0322
Carbon dioxide	3.59	0.0427
Carbon monoxide	1.49	0.0399
Chlorine	6.49	0.0562
Ethane	5.49	0.0638
Ethylene	4.47	0.0571
Helium	0.034	0.0237
Hydrogen	0.024	0.0266
Hydrogen chloride	3.67	0.0408
Hydrogen bromide	4.45	0.0433
Methane	2.25	0.0428
Neon	0.21	0.0171
Nitric oxide	1.34	0.0279
Nitrogen	1.39	0.0319
Oxygen	1.36	0.0318
Sulphur dioxide	6.71	0.0564
Water	5.46	0.0305

The constant a may be regarded as *measure of the van der Waals forces of cohesion existing* between the molecules of a given gas. The greater the value of a, the greater is the strength of the van der Waals forces. The values of a for hydrogen and helium are very small, being 0.024 and 0.034 respectively, indicating that the cohesive forces (van der Waals forces) in these gases are very weak.

The values of a for ammonia, carbon dioxide, chlorine and sulphur dioxide are very high being 4.17, 3.59, 6.49 and 6.71, respectively. This reveals that van der Waals forces in these gases are very strong. The greater the value of a, the greater would be the ease with which a gas can be liquefied.

EXPLANATION OF BEHAVIOUR OF REAL GASES O THE BASIS OF VAN DER WAALS EQUATION

$$\left(P + \frac{a}{V^2}\right)(V - b) = RT \quad \text{...van der Waals equation of state}$$

or $$PV + \frac{a}{V} - Pb + \frac{ab}{V^2} = RT$$

On Neglecting $\frac{ab}{V^2}$ a product of two very small quantities, we obtain

$$PV - + \frac{a}{V} = RT = P'V'$$

where P′ and V′ refer to the ideal pressure and volume of the gas respectively.

(i) At low pressure: Small addition of pressure due to mutual attraction plays an important part while b is negligible as compared to large total volume V. The factor a/V, therefore overweights Pb *i.e.*,

$$\frac{a}{V} > Pb$$

$$\frac{a}{V} - Pb = Z, \text{ a positive quantity.}$$

From eq. (1), we have

$$PV + Z = RT = P'V'$$

$$PV = RT - Z = P'V' - Z$$

i.e., the observed PV is less than P′V′.

The observed product PV at low pressure would be less than P′V′, the product of pressure and volume if the gases were ideal.

At low pressures, therefore PV decreases with increase in pressure in case of N_2, O_2 and CO_2, etc.

(ii) At high pressure: If pressure is high, the volume is small and small volume occupied by molecules themselves cannot be neglected. Whereas in comparison with the high pressure, small addition of pressure due to mutual attraction is negligible. Hence Pb overweights a/V, *i.e.*,

$$Pb > \frac{a}{V}.$$

i.e., $\frac{a}{V} - Pb = -y$, a negative quantity

$$\therefore \qquad PV + (-y) = P'V'$$

$$PV - y = P'V'$$

$$PV = P'V' + y$$

i.e., the observed PV is greater than the value expected from an ideal gas.

This explains the fact that PVl, after reaching a minimum, increases with further increase in pressure.

(iii) At ordinary (medium) pressure: It is seen that at ordinary temperature the factor $\frac{a}{V^2}$ becomes predominant at low pressures whereas the factor Pb predominates at high pressures. Therefore, these two factors will balance each other at some intermediate range of pressures in which the gas will show the ideal behaviour.

Hence at medium pressure, $Pb = \frac{a}{V}$

The equation (1) is reduced to

$$PV = RT = P'V'$$

and the gas behaves like an ideal gas.

(iv) At high temperature: At high temperatures, V is very large and hence both Pb and $\frac{a}{V}$ will be negligibly small, so that we have

$$PV = RT = P'V'$$

i.e., at high temperatures the behaviour of a real gas nearly agrees with the expected for ideal gas.

This explains why the deviations become less at high temperature.

(v) At low temperature: Both P and V are small, hence both pressure (attraction) and volume corrections are appreciable, with the result that the deviations are more pronounced.

(vi) Exceptional behaviour of hydrogen and helium: The molecules of these gases are of very small mass and consequently the force of attraction between the molecules is always negligible.

Hence neglecting the term $\frac{a}{V}$ equation (1) gets reduced to

$$PV - Pb = RT = P'V'$$

$$PV = P'V' + Pb$$

i.e., the product PV would become greater than P'V' event at low pressures.

EXPERIMENTAL DETERMINATION OF CRITICAL CONSTANTS

(A) Determination of Critical Temperature (T_c) and Critical Pressure (P_c)

Andrews determine critical temperature and critical pressure of a gas from its isotherms at different temperatures. This is a simple method which is based on the principle that *at critical temperature the surface of separation between liquid and vapour phases disappears.* This method is used when the substance is in liquid state at ordinary temperature.

The apparatus consists of a U-shaped hard glass tube having one limb smaller than the other. The tube is filled partially with mercury enclosing a small quantity of dry air in the longer limb. The upper portion of the longer tube is graduated. A small quantity of liquid and its vapour is kept over the mercury in the shorter limb and the mouth is sealed. The portion of to he shorter tube having liquid vapour, is then enclosed i a jacket, the temperature of which may be changed slowly and gradually when desired by circulating a suitable liquid from a thermostat.

The temperature of jacket is gradually increased. As the liquid is being heated, its surface tension is slowly decreasing and the boundary (or meniscus) would become progressively flatter and flatter. When the critical temperature is reached, the boundary or meniscus between the liquid and the vapour flickers and suddenly would disappear. The temperature at which this takes place is noted. The tube is then cooled a bit slowly and temperature is again noted when the gas becomes turbid for a moment and a small amount of liquid quickly settles at the bottom with the subsequent reappearance of the meniscus. The mean of the two temperature radians is regarded to be the critical temperature (T_c). At the critical temperature the reading of the corresponding pressure from the manometer is also noted. The mean of the two pressure reading corresponding to the two temperatures would five the critical pressure (P_c).

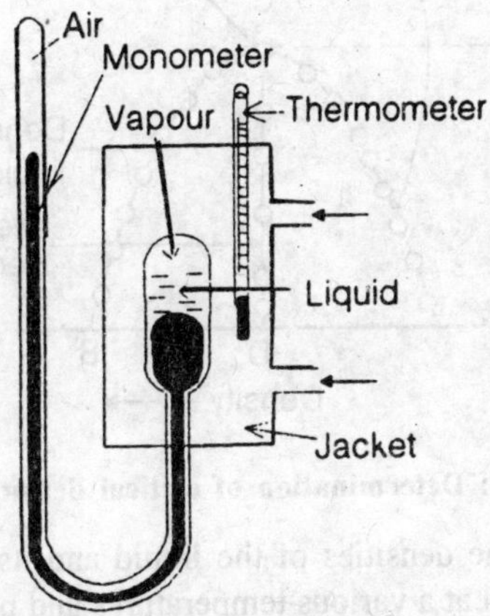

Fig. 4.7 : Cagniard de la Tour's apparatus.

B. Determination of Critical Volume (V_c)

It is not possible to determine the critical volume (V_c) of the gas experimentally because slight changes in pressure and temperature of the gas give rise to large changes in volume. So it becomes difficult to determine the V_c accurately by experiment.

However, the critical volume (V_c) can be determined indirectly by using the well known rule of rectilinear diameter due to Cailletet Mathias known as Cailletet and mathias rule which states that *the mean value of densities of any substance in the liquid state and its saturated vapour at the same temperature is a linear function of temperature i.e., the mean of the densities of liquid and its saturated vapour would be lying on a straight line called the rectilinear diameter.*

Thus if the density of a liquid is plotted against temperature, a curve of the type BC (Fig. 4.8) is obtained. If the density of the saturated vapour of the liquid is plotted against temperature, curve AC is obtained. The two curves meet at a point C which must correspond to the density at critical temperature. At point C, the liquid and its vapour become co-existent and identical and this point must give the critical density. In practice some difficulty arises in getting the upper portions of the two curves as the conditions existing there are uncertain. Hence, the upper portions of the curves would be almost approximate.

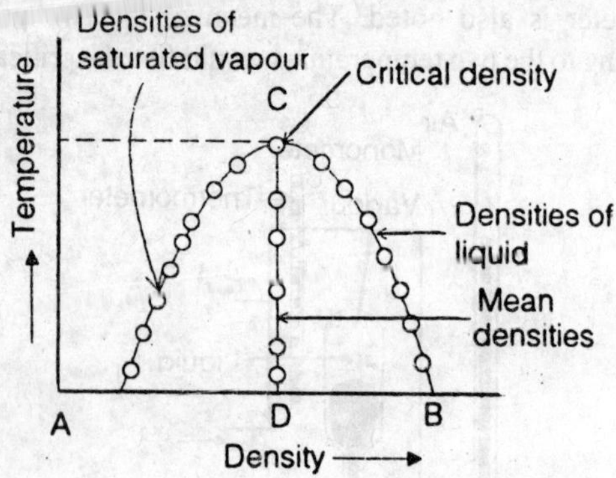

Fig. 4.8 : Determination of critical density (Dc).

Consequently the densities of the liquid and its saturated vapours would be determined at a various temperatures and plotted against T up to a few degrees below T_c (critical temperature). The curve is then made to complete. The mean densities are then plotted against the various temperatures (T) when a rectilinear (straight) line DC is obtained. The rectilinear diameter is then made to extend to meet the curve at C. The point C, therefore, corresponds to T_c (critical temperature) and critical density (D_c) of the substance. The value V_c is then obtained as follows:

$$\text{Critical volume } (V_c) = \frac{\text{Mol. wt. of sustance}}{\text{Critical density}} = \frac{M}{D_c}$$

Critical constants of some common gases are give in table 3.

Table 3 : Critical constants of gases.

Gas	*Critical Temperature, °K*	*Critical Pressure, atm*	*Critical Volume, ml.*
Helium, He	5.2	2.26	60.68
Hydrogen, H_2	33.2	12.28	64.51
Nitrogen, N_2	126.0	33.5	90.03
Carbon monoxide, CO	134.4	34.6	90.03

Argon, Ar	150.7	48.2	78.43
Oxygen, O_2	154.3	49.7	74.42
Methane, CH_4	190.2	45.6	98.76
Carbon dioxide, CO_2	304.3	72.2	95.65
Nitrous oxide, N_2O	306.9	71.9	96.91
Hydrochloric acid, HCl	324.5	81.6	89.90
Hydrogen sulphide, H_2S	373.5	89.0	126.86
Ammonia, NH_3	405.5	111.5	72.03
Chlorine, Cl_2	417.1	76.1	123.91
Sulphur dioxide, SO_2	430.3	77.6	124.75

The P-V Isotherms of Carbon Dioxide—Critical Phenomenon

The curves representing the variation of volume and pressure at constant temperature are termed as Isotherms. (Greek : isos = 'equal'). In the case of an ideal gas the isotherms are almost rectangular hyperbolas (Fig. 4.9).

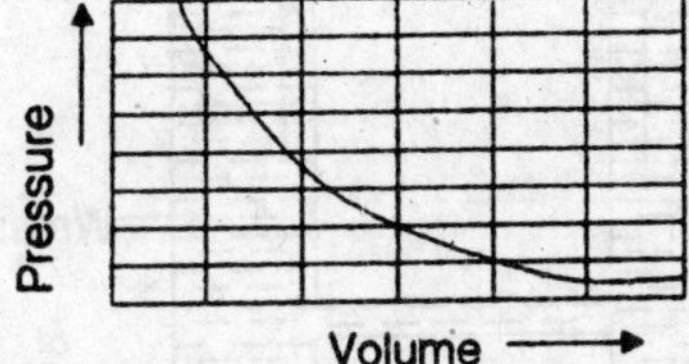

Fig. 4.9 : Rectangular hyperbola—isotherm in case of an ideal gas.

In 1869 T. Andrews attempted to study the effect of pressure on volume of CO_2 at different temperatures. In this experiment, some CO_2 gas was enclosed in a capillary tube and the volume occupied by the gas was measured at different pressures upto about 400 atmospheres.

His apparatus consisted of a sort of hydraulic press as shown in Fig. 4.10. Two similar capillaries were taken. One was filled partially with carbon dioxide and a pellet of mercury was introduced in the tube to act as a seal. In the other capillary, air was filled and sealed by mercury. These capillaries were then fixed into metal compression chambers by liquid tight joint. The chambers were interconnected ad carried screw plugers at the bottom to apply pressure. The upper portions of the tubes were surrounded by water bath to keep the temperature constant.

If the volume readings were plotted against the pressure a set of curves of the type shown in Fig. 4.11 were obtained. Each curve would correspond to one particular temperature and is, therefore, called isotherm. The curves should be rectangular hyperbolas like that expected for an ideal gas. It is quite evident that CO_2 does not yield rectangular hyperbolas although with increasing temperature the curves approximate to one. We will now consider the various isotherms one by one.

(a) *At 13.1°C:* From the isotherm it can be seen that carbon dioxide is entirely gaseous at low pressures (*e.g.*, along AB) ad the volume decreases with increasing pressure in the normal fashion (*i.e.*, AB is roughly hyperboloid). The point A would correspond to gaseous CO_2 occupying a certain volume under a certain pressure. At low pressures, if the pressure gets increased, the volume would decrease as shown by the curve AB.

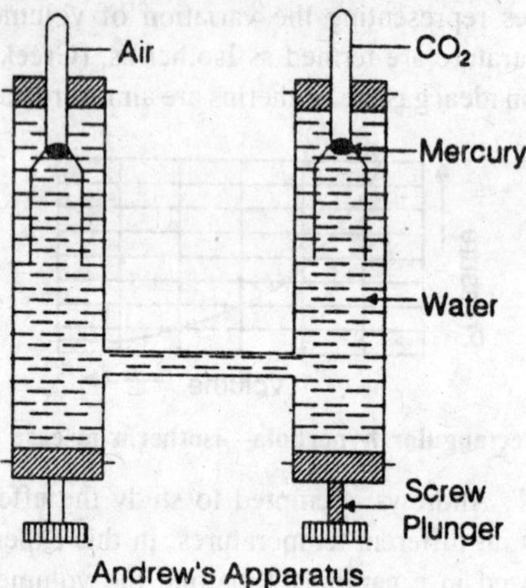

Fig. 4.10 : Andrew's apparatus.

At B liquefaction starts; vapour gets converted into liquid of very much greater density and the volume gets decreased suddenly as the volume occupied by a liquid would be much less than that of gas so that BC would be almost parallel to the volume axis-indicating the sudden decrease in volume occurring at the same pressure (49.8 atmospheres) as more and more of the gas changes into liquid.

Liquefaction gets completed at C and CD reveals the small effect of pressure on a liquid. The steep line CD which is almost vertical represents the compressibility of a liquid and reveals that the liquids are much less compressible, *i.e.*, increase in pressure will produce a very small decrease in volume.

In order to summaries, along curve AB there is only vapour, along CD only liquid and along BC liquid and vapour co-exist. As BC is parallel to the volume axis, there can be only one pressure at which liquid and vapour co-exist. This pressure is termed as the *vapour pressure* of the gas.

(b) The next isotherm EFGH at 21.1°C is also similar with the difference that the horizontal portion where liquefaction occurs is slightly short (*i.e.*, the range of volumes over which gas and liquid can co-exist is smaller) and the vapour pressure higher (*i.e.*, the pressure at which liquefaction commences is higher).

(c) As the temperature gets raised, this horizontal portion would become less and less till at 31.1°C it reduces almost to a point X. Above this temperature there occurs no trace of horizontal portion, *i.e.*, beyond 31.1°C there occurs no indication of liquefaction at all.

Anderws applied a pressure of 300 to 400 atmospheres but could to liquefy CO_2 if the temperature was beyond 31.1°C. Below 31.1°C a pressure of 75 atmospheres was quite sufficient to liquefy carbon dioxide.

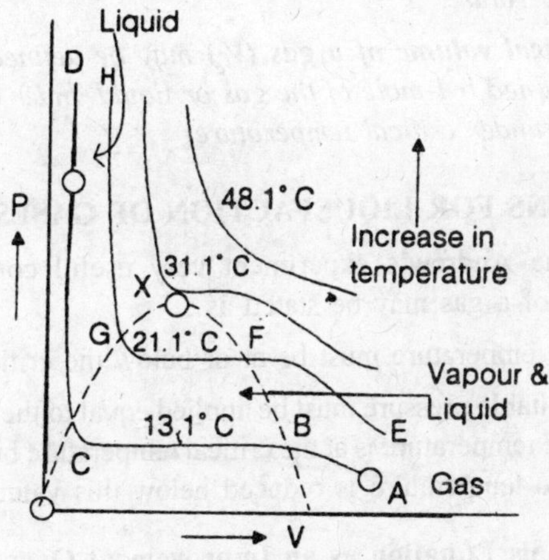

Fig. 4.11 : Isotherms of CO_2 near the critical temperature.

Point X refers to the critical point for CO_2. The distinction between liquid and vapour state disappears at this point and CO_2 exists in a state called the critical state. The isotherm passing through critical point is termed as the critical isotherm.

From the experiment Andrews was able to conclude that if the temperature of CO_2 is above 31.1°C it cannot be liquefied no matter how high the pressure may be. This maximum temperature (31.1°C) at which liquefaction can be brought about is termed as critical temperature.

The pressure required to liquefy a gas at the critical temperature is termed as the critical pressure. If the temperature is below the critical temperature, the gas will liquefy at a somewhat lower pressure. All other gases have been found to behave similarly. The critical phenomenon observed b Andrews for CO_2 can be observed for any gas.

CRITICAL CONSTANTS

(i) ***Critical temperature of a gas (T_c)*** *may be defined as the minimum temperature which must be reached before it can be liquefied by the application of pressure.* A gas cannot be liquefied by pressure above its critical temperature, no matter how high the applied pressure may be. Each gas is having definite critical temperature, *e.g.*, critical temperature of CO_2 is 31.1MC.

(ii) ***Critical pressure of a gas (P_c)*** *may be defined as the minimum pressure necessary for the liquefaction of a gas at critical temperature.*

(iii) ***Critical volume of a gas (V_c)*** *may be defined as the volume occupied b 1 mole of the gas or liquid under critical pressure and under critical temperature.*

CONDITIONS FOR LIQUEFACTION OF GASES

From the Andrew's experiment very useful conditions for the liquefaction of a gas may be stated as :

(i) The temperature must be at or below the critical temperature.

(ii) A suitable pressure must be applied equal to the critical pressure if the temperature is at the critical temperature but becoming less if the temperature is reduced below this value.

van der Waals Equation as an Improvement Over Simple Laws

Van der Waals equation is an improvement over simple laws due to following reasons:

(i) *Van der Waals equation explains the devication from Boyle's law:* Regnault and Amagat studied the behaviour of cases and observed that they do not obey simple gas laws under different conditions of temperature and pressure, more particularly at low temperature and high pressure. They also plotted graphs between the product of pressure and volume ad pressure and found that PV product was not constant as desired by Bole's law. These deviations from ideal behaviour were explained by van der Waals equations.

(ii) *It also explains critical phenomenon.* This equation satisfactorily explains critical phenomenon, observed in the case of gases. The theoretical isotherm (*i.e.*, the curves between V and P at constant T) obtained from the values of V and the experimental values of 'a' and 'b' from van der Waals equation were found to be quite similar. The theoretical isotherms obtained from simple gas laws are quite different from the experimental isotherms.

PRINCIPLE OF CONTINUITY OF STATES

Although, the transition from liquid to vapour and *vice versa* generally takes place suddenly and the two states can be sharply distinguished from each other as can be see from AB and CD parts of 13.1°C isotherm for carbon dioxide (Fig. 4.11), yet where occurs a possibility of passing from one state of aggregation to the other without involving sudden change.

Hence for further theoretical considerations it becomes essential to show that the liquid stated does not refer to a sharp and discontinuous transition from gaseous state but rather a continuation of the gaseous phase into the region of very strong intermolecular attractions where the volume becomes very small. The idea of a contiguity from liquid to gaseous state will become evident from the following consideration:

The dome-shaped curve depicted b the dashed line in Fig. 4.11 connects the end points of the horizontal portions of the isotherms.

As a horizontal portion represents two phases in equilibrium, all points inside the dome correspond to equilibrium between liquid and vapour. For such points it always becomes possible to draw any sharp distinction between liquid and vapour, because there is a definite boundary surface between the two phases.

For points lying outside the dome only there is only one phase (liquid or vapour) ad here it is not possible to draw any sharp distinction between liquid and vapour. This fact is termed as the *principle of contiguity of states* and can be demonstrated with reference to Fig. 4.11.

Suppose, that carbon dioxide gas under atmospheric pressure is made to heat a temperature above the critical temperature, as to 48°C and that is then subjected to a pressure greater than critical pressure, as, to 100 atmospheres, the system will change volume as shown by the isotherm 48°C.

When the temperature is lowered to 13.1°C while maintaining the pressure at 100 atmospheres, a point on the curve CD will be reached a curve which applies to liquid carbon dioxide. At no point in the process there is any abrupt change of state; the change from gas to liquid will be contiguous. Similarly, it is possible to start with liquid carbon dioxide under the conditions represented, say by the point C.

If we now raise the pressure o the liquid to 10 atmospheres and the temperature to 48°C the condition of the system will be represented by a point on that isotherm; and on reducing the pressure to that of the atmosphere ad then allowing the temperature to fall to 13.1°C, we shall be having carbon dioxide as a gas; and at no time during the whole process any discontinuity of properties will be observed, or an separation into liquid and vapour. The passage from the liquid to the gaseous state would be continuous.

An important conclusion derived from the principle of continuity of states is that one may think of a liquid near the critical point as just a highly compressed gas. If a equation of state can be found which accurately describes the gas close to its critical temperature and pressure that equation should also be able to describe the liquid.

VAN DER WAALS EQUATIONS AND CRITICAL STATE—RELATIONS BETWEEN VAN DER WAALS CONSTANTS AND CRITICAL CONSTANTS

A very good confirmation of the general character of van der Waals equation may be provided b its application to critical phenomenon.

The van der Waals equation may be put as follows:

$$\left(P + \frac{a}{V^2}\right)(V - b) = RT \text{ (putting } = 1) \qquad ...(1)$$

$$PV + \frac{a}{v} - Pb - \frac{ab}{V^2} = RT$$

On multiplying throughout by V^2, we obtain

$$PV^3 + aV - PbV^2 - ab = RTV^2$$

or $$PV^3 + aV - PbV^2 - ab - RTV^2 = 0$$

On arranging in descending powers of V, we get

$$PV^3 - PBV^2 - RTV^2 + aV - ab = 0$$

On dividing b P, we get

$$V^3 - bV^2 - \frac{RTV^2}{P} + \frac{aV}{P} - \frac{ab}{P} = 0$$

$$V^3 - V^2 - \left(b + \frac{RT}{P}\right) + \frac{aV}{P} - \frac{ab}{P} = 0 \qquad ...(2)$$

This is a third degree equation in terms of V (cubic equation in V) and below a certain value of temperature T the equation can be having one of three real roots. With the rise of temperature these roots would approach one another and become identical ultimately.

This can be readily understood if a theoretical P – V graph is plotted from the van der Waals equation. The curves obtained are of the type ABEFGCD as depicted in Fig. 4.12. The three real unequal roots correspond to point B, F and C. The points B and C represent saturated vapour and liquid respectively, but F is having no physical significance.

As the temperature gets raised the S-shaped dotted portion of the curve gets reduced to a point. The values for the volume of liquid and the gas approach each other until at this volume of liquid and the gas approach each other until at this point, the three values of V become identical. This point would correspond to the critical temperature T_c and the volume V_c.

At the critical point, the three roots of the van der Waals equation would become identical and equal to V_c. But this refers to the value of the three roots of the van der Waals equation at the critical point.

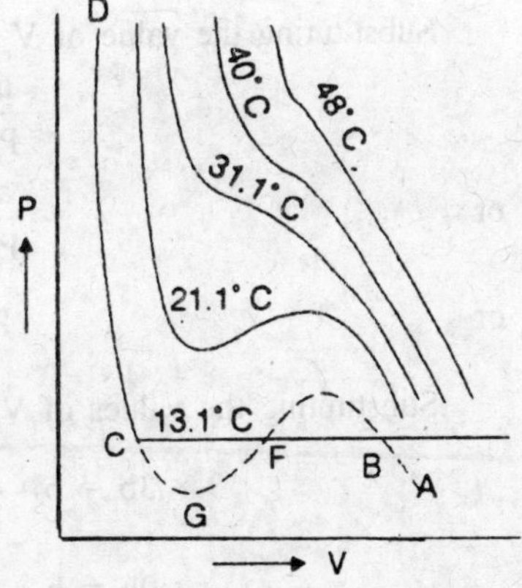

Fig. 4.12 : Isothermals of CO_2 from Van der Waals equation.

$$(V - V_c)^3 = 0 \qquad ...(3)$$

or $$V^3 - 3V_cV^2 + 3V_c^2V - V_c^3 = 0 \qquad ...(4)$$

On substituting these values in eq. (2), we get,

$$V^3 - V^2\left(b + \frac{RT_c}{P_c}\right) + \frac{aV}{P_c} - \frac{ab}{P_c} = 0 \qquad ...(5)$$

As both eqs. (4) and (5) are identical then the coefficients of the equal powers of V i the two equations may be equated:

$$3V_c = b + \frac{RT_c}{P_c} \qquad ...(6)$$

$$3V_c^2 = \frac{a}{P_c} \qquad ...(7)$$

$$V_c^3 = \frac{ab}{P_c} \qquad ...(8)$$

On dividing Eq. (8) by Eq. (7), we get,

$$\frac{V_c^3}{3V_c^2} = \frac{ab}{P_c} \times \frac{P_c}{a}$$

or $$\frac{V_c}{3} = b$$

$$V_c = 3b \qquad ...(9)$$

Substituting the value of V_c from Eq. (9) into Eq. (7), we obtain

$$\frac{a}{P_c} = 3 \times (3b)^2$$

or $$\frac{a}{P_c} = 27b^2$$

or $$P_c = \frac{a}{27b^2} \qquad ...(10)$$

Substituting the values of V_c and P_c in equation (6), we get

$$3 \times 3b = b + \frac{RT_c}{a/27b^2}$$

$$9b = b + \frac{RT_c}{a/27b^2}$$

or $$8b = RT_c \times \frac{27b^2}{a}$$

or $$T_c = 8b \times \frac{a}{27b^2} \times \frac{1}{R}$$

$$T_c = \frac{8a}{27bR} \qquad ...(11)$$

Derivation of $P_cV_c = \frac{3}{8}RT_c$ from van der Waals Equation and calculation of van der Waals constants 'a' and 'b' in terms of T_c and P_c.

From van der Waals equation, we get

$$V^3 - V^2\left(b + \frac{RT_c}{P_c}\right) + V\frac{a}{P_c} - \frac{ab}{P_c} = 0 \qquad ...(12)$$

and $$V^3 - 3V_cV^2 + 3V_c^2V - V_c^3 = 0 \qquad ...(13)$$

On equating the coefficients of (12), and (13), we get

$$3V_c = \frac{RT_c}{P_c} + b \qquad ...(14)$$

$$3V_c^2 = \frac{a}{P_c} \qquad ...(15)$$

$$V_c^3 = \frac{ab}{P_c} \qquad ...(16)$$

On dividing Eq. (16) by (15), we get

$$V_c = 3b \quad \text{or} \quad b = \frac{1}{3}V_c \qquad ...(17)$$

On substituting the value of b in (14), we obtain

$$3V_c = \frac{RT_c}{P_c} + \frac{V_c}{3}$$

or $$3V_c - \frac{V_c}{3} = \frac{RT_c}{P_c}$$

or $$\frac{8}{3}V_c = \frac{RT_c}{P_c}$$

or $$P_cV_c = \frac{3}{8}RT_c \qquad ...(18)$$

On rearranging Eq. (18), we get

$$V_c = \frac{3}{8}\frac{RT_c}{P_c}$$

On squaring, we obtain

$$V_c^2 = \frac{9}{64}\frac{R^2T_c^2}{P_c^2} \qquad ...(19)$$

On substituting the value of $V_c^2 = \frac{a}{3P_c}$ from Eq. (15) into Eq. (19), we get,

$$\frac{a}{3P_c} = \frac{9}{64}\frac{R^2T_c^2}{P_c^2}$$

$$a = \frac{27R^2T_c^2}{64P_c} \qquad ...(20)$$

On substituting the value of $V_c = 3b$ in Eq. (14), we obtain,

$$9b = \frac{RT_c}{P_c} + b$$

or

$$8b = \frac{RT_c}{P_c}$$

or

$$b = \frac{RT_c}{8P_c} \qquad ...(21)$$

THE PRINCIPLE OF CORRESPONDING STATES

In 1881, van der Waals demonstrated that if the pressure volume and temperature of a gas are expressed in terms of its critical pressure critical volume and critical temperature, then an important generalization called the principle of corresponding states would be obtained.

According to van der Waals equation, we have

$$\left(P + \frac{a}{V^2}\right)(V - b) = RT \qquad ...(1)$$

This equation has constants 'a' and 'b' which cannot be measured directly. Therefore, this equation is not applicable to all gases and in order to make it so, these constants have to be removed.

It has been found experimentally that near the critical point all gases give the same type of isotherms as carbon dioxide, but the actual position of the critical isotherms gets varied with the critical constant of the gas studied. The similarity is great enough to postulate that if for each different gas a suitable scale of pressure, volume and temperature is selected, it will be possible to superimpose all the critical isotherms on one another. If this were done of the critical isotherms it becomes probable that all other isotherms would coincide.

Under such circumstances, there should be one isothermal diagram and one fundamental equation for all gases. Such a scale will be one in which the pressure, volume, and temperature will be expressed as fractions of the critical pressure, volume and temperature.

In order to obtain such an equation, the values of P, V and T must be expressed as corresponding fractions of P_c, V_c and T_c, the critical pressure, volume and temperature,

$$\pi = P/P_c \quad \text{or} \quad P = \pi P_c$$

$$\phi = V/V_c \quad \text{or} \quad V = \phi V_c$$

$$\theta = T/T_c \quad \text{or} \quad T = \theta T_c$$

where π, ϕ and θ are referred to as *reduced pressure, reduced volume and reduced temperature respectively.*

On substituting the values of P, V and T in the van der Waals equation for one mole of an ideal gas, we get

$$\left(\pi P_c + \frac{a}{\phi^2 V_c^2}\right)(\phi V_c - b) = R_q T_c \qquad ...(2)$$

$$P_c = \frac{a}{27b^2}, \quad V_c = 3b$$

$$\text{and} \quad T_c = \frac{8a}{27Rb}$$

On substituting these values in equation (2), we get,

$$\left(\pi \frac{a}{27b^2} + \frac{a}{9\phi^2 b^2}\right)(3\phi b - b) = R\theta \frac{8a}{27Rb}$$

or

$$\frac{a}{27b^2}\left(\pi + \frac{3}{\phi^2}\right) \times b(3\phi - 1) = \frac{a}{27b} 8\theta$$

or

$$\frac{a}{27b}\left(\pi + \frac{3}{\phi^2}\right)(3\phi - 1) = \frac{a}{27b} 8\theta$$

or

$$\left(\pi + \frac{3}{\phi^2}\right)(3\phi - 1) = 8\theta$$

This equation involving neither of the characteristic constants a, b, T_c, P_c and V_c is known as van der Waals reduced equation of state. This equation is applicable to all substances (liquid or gaseous) irrespective of their nature, because it is not involving either of the characteristic constants.

It follows from the above equation that *fi two substances are at the same reduced temperature* (θ) *ad pressure* (π) *they must have the same reduced volume* (ϕ). This statement is called the law of corresponding states or principle of corresponding states.

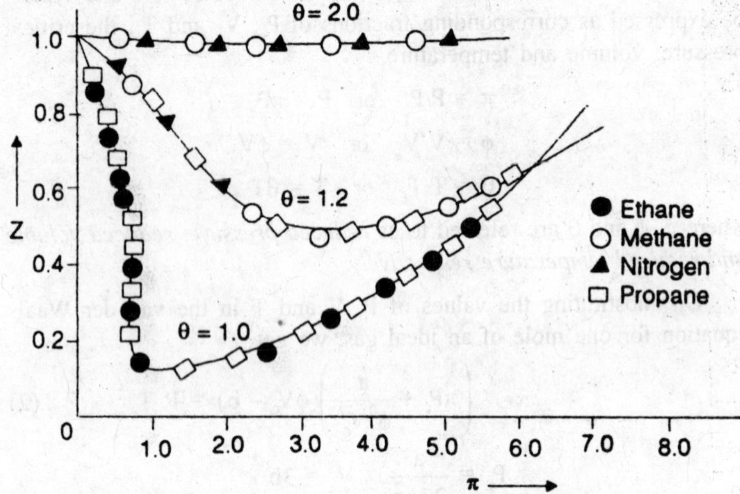

Fig. 4.13 : The compressibility factor as a function of reduced pressure and reduced temperature.

Significance of Law of Corresponding States

It has a great significance in the study of the relationship between physical properties ad chemical constitution of various liquids. As it is known that the pressure has practically no effect on liquids, so while studying the relation between physical properties and chemical constitution the determination of these properties should be performed at the same reduced temperature.

It has been found that the *boiling point of a liquid at atmospheric pressure is almost 2/3rd of its critical temperature* (T_c) *when expressed in absolute scale*. Consequently, boiling point is considered as the *corresponding temperature*. So, various physical properties of liquids such as molecular volumes of various liquids should be compared at their respective boiling points.

In actual practice one cannot use the above reduced equation of state directly. One makes use of the graphs between compressibility factor 'Z' and the reduced pressure at different reduced temperatures. The same graph is applicable to all gases. This can be seen from the following considerations, since $Z = \frac{PV}{RT}$.

Writing this in terms of reduced terms,

We have $$Z = \frac{(\pi P_c)(\phi V_c)}{R(\theta T_c)} = \left(\frac{P_c V_c}{RT_c}\right)\frac{\pi\phi}{\theta} = \frac{3}{8}\frac{\pi\phi}{\theta}$$

Thus, if two gases are having the same reduced temperature and reduced pressure, then their reduced volume will be the same, thus R.H.S. is independent of the as, thus all gases will have the same value of 'Z'. If isotherms are made to plot in terms of the reduced variables, the same curve should be obtained for all gases. The principle of corresponding states thereby enables the properties of real gases to be plotted o a single diagram (Fig. 4.13).

LIQUEFACTION OF GASES

For every known gas there is a temperature above which it is impossible to liquefy a gas, o matter how great the pressure may be. This temperature is known as the critical temperature. The minimum pressure necessary to liquefy a gas at its critical temperature is its critical pressure.

There are many gases such as ammonia, chlorine, sulphur dioxide, carbon dioxide, etc. which can be liquefied by the application of a suitable pressure alone because they are having high critical temperatures. On the other hand, there are gases such as hydrogen, oxygen, nitrogen and helium which cannot be liquefied by the simple application of a suitable pressure alone because they are having very low critical temperatures. These gases were regarded as *permanent* because the real significance of critical temperature was misunderstood for a longtime. But now it is known that these gases could be liquefied if they are first cooled below their respective critical temperatures and then made to compress.

The various methods used for the liquefaction of gases are as follows:

(1) *By the use of freezing mixture:* Faraday (1823) was able to liquefy a number of SO_2, Cl_2, CO_2, etc. by subjecting them to

high pressures at low temperatures. He used a V-shaped tube, in one arm of which the gas was prepared and i the other end it liquefied under its own pressure and with the help of external cooling (Fig. 4.14). But Faraday was not able to liquefy the gases like hydrogen, oxygen and nitrogen, even when the pressure was as high as 2790 atmospheres. These gases were, therefore, known as *permanent gases*.

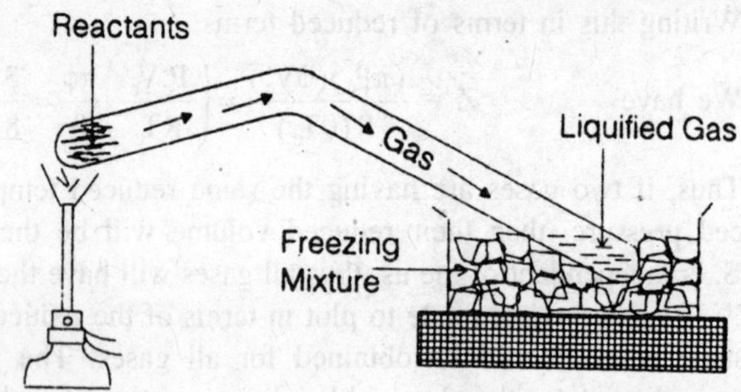

Fig. 4.14 : Faraday's method of liquefaction of gases.

In 1869, Andrews studied the effect of pressure on CO_2 at different temperatures and observed that for each gas there exists a temperature, called *critical temperature*, above which it is not possible to liquefy the gas by compression no matter how great may be the pressure. The ignorance of this fact was the main reason why the early attempts to liquefy the so-called 'permanent' gases got failed.

(2) *By the use of volatile refrigerants:* This method is based on the fact that when a volatile compound evaporates rapidly, it absorbs its latest heat of vaporisation from the surroundings and consequently temperature of the surroundings becomes low.

Pictet was successful in liquefying CO_2 by the cooling caused by the evaporation of liquid SO_2 around it. Thilosier (1834) succeeded in solidifying liquid CO_2 by allowing it to evaporate freely under atmospheric pressure. Thilosier, obtained temperature – 100 to – 110°C by allowing a mixture of a solid CO_2 (dry ice) and ether (called Thilosier's mixture) to evaporate under reduced pressure.

Faraday was able to liquefy most of the common gases by using Thilosier's mixture. He however failed to liquefy O_2, N_2, H_2, etc. and the called them permanent gases.

(3) *By the use of Joule-Thomson effect* (By expansion of cold compressed gases).

Joule-Thomson Effect

The English physicists, James Joule and William Thomson (later Lord Kelvin), observed that if a *gas under high pressure is allowed to expand into a region of low pressure it suffers a fall in temperature.* This phenomenon is termed as Joule Thomson effect.

Joule-Thomson effect *provides further support to the view that attractive forces do exist between gas molecules.* When the gas expands, the molecules fall apart from one another. Therefore, work has to be done so as to overcome the cohesive or attractive forces which tend to hold the molecules together. Hence, work is done by the system at the expense of the kinetic energy of the gaseous molecules. Consequently, the kinetic energy would decrease and since this is proportional to temperature, cooling results. It may be noted that in this case no external work is being done by the as in expansion.

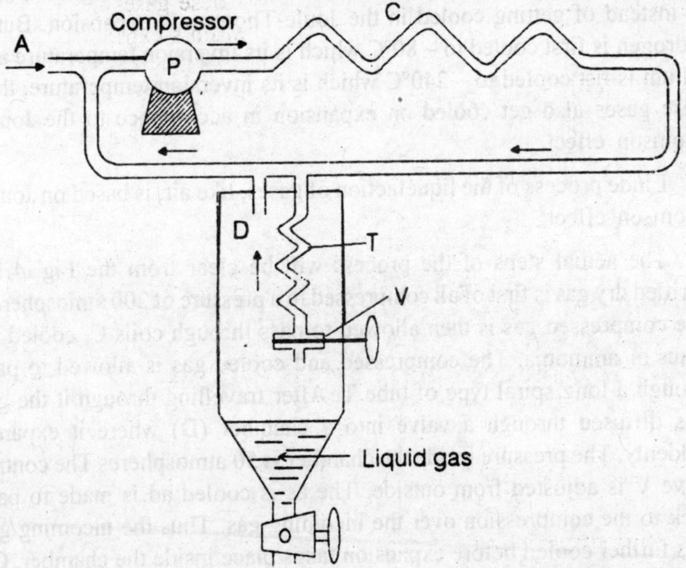

Fig. 4.15 : Linde's apparatus for liquefaction of gases.

Inversion Temperature

Experiment has revealed that gases become cooler during the Joule-Thomson's expansion *only if they are below a certain temperature called the* inversion temperature, T_i. The inversion temperature is characteristic of each gas. It is related to van der Waals constant a and b of the gas connected by the following relation:

$$T_i = \frac{2a}{Rb} \quad \text{...(1)}$$

At the inversion temperature, there occurs no Joule-Thomson effect.

Thus, if a gas under pressure is allowed to pass through a porous plug and expands adiabatically into a region of very low pressure *at the inversion temperature,* there occurs neither fall no rise in temperature. If, however, the expansion occurs *above* the inversion temperature, there occurs a small *rise* of temperature and if it occurs *below* the inversion temperature there occurs a small *fall* of temperature.

In most gases, the inversion temperature lies within the range of ordinary temperature. Hence, they are cooled in Joule-Thomson's expansion. Hydrogen and helium, however, have *very low inversion temperatures*. Thus, at ordinary temperatures, these gases get *warmed* up instead of getting cooled in the Joule-Thomson's expansion. But if hydrogen is first cooled to – 80°C which is its inversion temperature and helium is fist cooled to – 240°C which is its inversion temperature, then these gases also get cooled on expansion in accordance to the Joule-Thomson effect.

Linde process of the liquefaction of gases, like air, is based on Joule-Thomson effect.

The actual steps of the process will be clear from the Fig. 4.15. Purified dry gas is first of all compressed to a pressure of 200 atmospheres. The compressed gas is then allowed to pass through coils C, cooled by water or ammonia. The compressed and cooled gas is allowed to pass through a long spiral type of tube T. After travelling through it the gas gets diffused through a valve into a chamber (D) where it expands suddenly. The pressure inside the chamber is 50 atmospheres The control valve V is adjusted from outside. The as is cooled ad is made to pass back to the compression over the incoming gas. Thus the incoming gas gets further cooled before expansion takes place inside the chamber. On repeating the cycle several times, the temperature of the expanding gas would go sufficiently down ad a part of it gets condensed. The liquid collects at the bottom of the chamber from where it is drawn off. The whole arrangement is insulated by wool or left to avoid heating of the tubes.

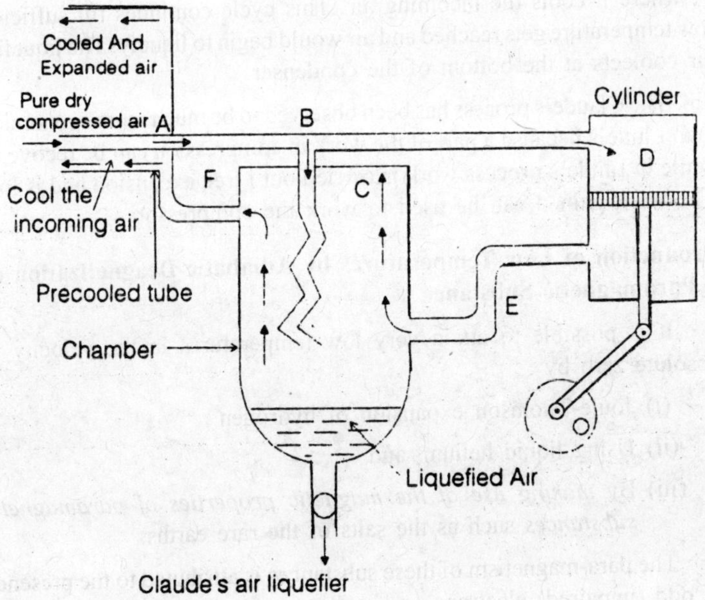

Fig. 4.16

By Adiabatic Expansion

If a gas is allowed to expand against an external pressure (say against a piston of an engine), it does some work. If a system is closed and insulated so that neither heat could enter the system nor go out of it, then the expansion would be almost adiabatic. This work done against the external pressure would be at the expanse of kinetic energy of the gas molecule. Consequently, a fall in temperature takes place. This principal combined with Joule-Thomson effect finds use in Claude's process for the liquefaction of air (Fig. 4.16).

Air free from CO_2 and moisture is made to compress to 40 - 50 atmospheres and then cooled by passing through a precooler. The gas then splits into two steams at B. A pat of the air enters the condenser C directly, where the other part enters the cylinder D where it undergoes expansion adiabatically, whereby the temperature of the gas falls. The cooled gas then enters the liquefying chamber via the side tube E.

As this cooled gas goes up in the chamber, it cools the air passing down the Jet J whereby further cooling takes place because of Joule-Thomson effect. The cooled gas then goes out through the precooler tube

T, where it cools the incoming air. This cycle continues till sufficient low temperature gets reached and air would begin to liquefy. The liquefied air coolects at the bottom of the condenser.

The Claude's process has been observed to be much more economical than Linde's because a part of the work in compression can be recovered while in Linde's process work is carried out i free expansion and is lost. The work gained can be used to work the compressors.

Production of Low Temperatures by Adiabatic Deagnetization of a Paramagnetic Substance

It is possible to attain very low temperatures in the velocity of absolute zero by :

(i) Joule-Thomson expansion of hydrogen

(ii) Using liquid helium, and

(iii) By *making use of the magnetic properties of paramagnetic substances* such as the salts of the rare earths.

The para-magnetism of these substances is attributed to the presence of odd (unpaired) electrons.

The method of adiabatic demonetization of a paramagnetic substance was first of all suggested i 1926 by Debye and independently i 1927 by Giauque. This method is based on the fact that *the temperature of thermally isolated paramagnetic substance drop shwne jit gets demonetized.* This can be best understood by considering the following equation for a magnetic system:

$$dS = \frac{C_b dT}{T} + \varepsilon_0 V \left(\frac{\partial M}{\partial T}\right)_B dB \qquad ...(1)$$

where M refers to the magnetization in the presence of the external magnetic field B; V the volume, ε_0 the vacuum permeability and C_B the heat capacity of the system at constant magnetic field.

The magnetization of a paramagnetic substance at constant magnetic field B always decreases with increase in temperature T so that the partial derivative $(\partial M/\partial T)_B$ is always negative. Hence, when the substance is demagnetized reversibly ad adiabatically (so that dB is negative and dS = 0), we get

$$\frac{dT}{T} = -\frac{\varepsilon_0 V}{C_B}\left(\frac{\partial M}{\partial T}\right)_B dB \qquad ...(2)$$

or $$\left(\frac{\partial T}{\partial B}\right)_S = \frac{\varepsilon_0 TV}{C_B}\left(\frac{\partial M}{\partial T}\right)_B \quad ...(3)$$

Eq. (3) is known to form the basis of the principle of adiabatic demagnetization which gives the rate of change of temperature with magnetic field upon reversible adiabatic demagnetization. It can be seen from Eq. (2) that dT/T is negative so that dT is also negative. Hence a paramagnetic substance would cool upon adiabatic demagnetization.

The temperature-entropy (T-S) curve for a paramagnetic salt, such as gadolinium sulphate octahydrate, is depicted in Fig. 4.17.

From the curve it can be seen that in the unmagnetized state (when B = 0), the spins of the unpaired electrons of the paramagnetic material get oriented in a random manner in all directions. Thus, there occurs a disorder.

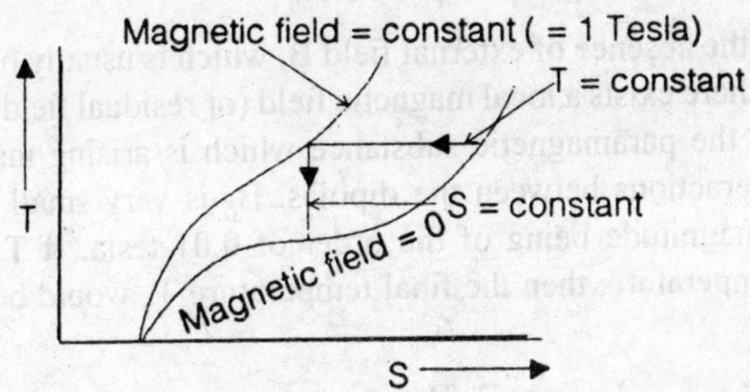

Fig. 4.17 : The T-S curve for a paramagnetic salt.

If, however, the substance is made to magnetize, the spins, and hence the magnetic moments associated with them, tend to align themselves with the magnetic field, thus reducing disorder. In other words, it means that the unmagnetized substance is having greater entropy than the magnetized substance at a given temperature. This can be seen from Fig. 4.17.

In the first stage of the process, the paramagnetic substance, in contact with a liquid helium bath at 1 K, would be *isothermally magnetized.* Under isothermal conditions (T = constant, dT = 0), Eq. (1) gets reduced to

$$TdS = \mu_0 TV(\partial M/\partial T)_2 dB \quad ...(4)$$

As now dB is positive and $(\partial M/\partial T)_B$ is always negative, an increase in B causes an overflow of heat and hence a *decrease* in entropy. This is exhibited by the T= constant line on the T-S curve of Fig. 17.

In the second stage, the substance is removed and *adiabatically demagnetized.* Now it cools. This is depicted by the S = constant line on the T-S curve. The extent of cooling has been found to depend upon the strength of the applied magnetic field, the intensity of magnetization and the heat capacity of the paramagnetic salt.

The experimental set-up used for carrying out adiabatic demagnetization is shown in Fig. 18. The paramagnetic substance A is made to suspend inside a cryostat which gets cooled by liquid helium boiling under reduced pressure. During the first isothermal stage, the space D is having helium gas; this space will permit heat exchange between the substance and liquid helium. During the second stage, the space D gets evacuate to establish adiabatic conditions. The cycle of isothermal magetization, followed by adiabatic demagnetization, gets repeated to get millidegree temperatures.

Even in the absence of external field B, which is usually of the order of 1-4 tesla, there exists a local magnetic field (or residual field) B_0 which is present in the paramagnetic substance which is arising mainly from magnetic interactions between the dipoles. B_0 is very small compared with B, its magnitude being of the order of 0.01 tesla. If T_1 refers to the initial temperature, then the final temperature T_2 would be given by the formula.

$$T_2/T_1 = B_0/B$$

If $T_1 = 1$ K,

$B_0 = 10^{-2}$ tesla, $B = 1$ tesla, then $T^2 \approx 10^{-2}$ K

We can remark here that the process of adiabatic demagnetization is also applicable to the magnetic moments of *nuclei.* The application of adiabatic demagnetization to nuclear magnetic moments has allowed a temperature of 10^{-6} K to be reached. However, one should remember that *it would not be possible to attain absolute zero of temperature by using a succession of adiabatic cooling processes.* This can be understood from the T-S curve of Fig. 4.18.

In 1949 W.F. Giauque won nobel prize in Chemistry for his contributions to thermodynamics and the production of low temperatures by adiabatic demagnetisation of a paramagnetic salt.

INTERMOLECULAR FORCES OR VAN DER WAALS FORCES

It is now well established that forces of attraction exist between *neutral molecules* as well. These are termed as intermolecular forces or cohesive forces or van der Waals forces. These forces arise from three in types of interactions, viz.,

(1) *Dipole-Dipole Interaction*

(2) Dipole-Induced Dipole Interaction, and

(3) Induced Dipole-Induced Dipole Interaction.

1. Dipole—Dipole Interaction

The *polar molecules* which, though neutral, are having permanent dipoles. In such molecules the van der Waals forces are mainly attributed to electrical interaction between the dipoles known as *dipole-dipole interaction*. For instance, gases like ammonia, sulphur dioxide, hydrogen fluoride, hydrogen chloride, etc., are having permanent dipoles as a result of which appreciable dipole-dipole interaction exists between the molecules of these cases.

The magnitude of this type of interacting has been found to depend upon the dipole moment of the molecule concerned. The greater the dipole moment the greater would be the dipole-dipole interaction. Due to the attractive iteraction, these gases can be readily liquefied.

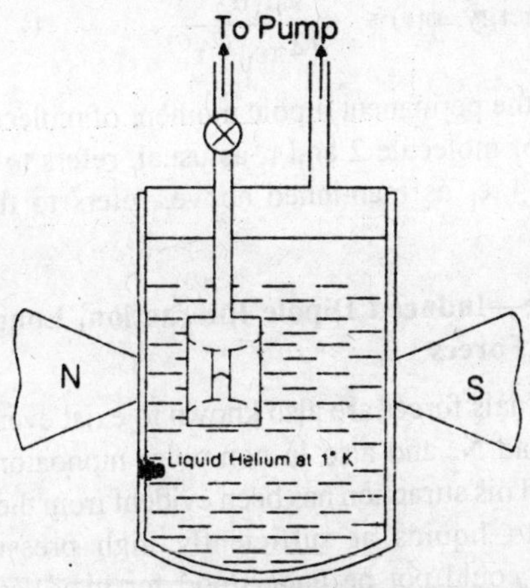

Fig. 4.18 : Experimental set-up for adiabatic demagnetization of a paramagnetic salt.

The average interaction energy of the two molecules with permanent dipole moments μ_1 and μ_2 is given by the expression

Interaction energy,

$$\phi(r) = -2\left(\frac{\mu_1\mu_2}{4\pi\varepsilon_0}\right)^2\left(\frac{1}{6r}\right)\left(\frac{1}{3kT}\right) \quad ...(1)$$

where r refers to their separation, k Boltzmann constant and $4\pi\varepsilon_0$ the permittivity factor of the vacuum.

2. Dipole—Induced Dipole Interaction

A polar molecule may some times be able to polarise a neutral molecule which lies in its vicinity and thus brings about dipolarity i that molecule just as a magnet is able to induce magnetic dipolarity i a neutral piece of iron lying close by. The induced dipole then undergoes interaction with the dipole moment of the first molecule and thereby the two molecules get attracted together. The magnitude of this interaction would, evidently, depend upon the magnitude of the dipole moment of the polar molecule and the polarizability of the neutral molecule. The average interaction energy for the two molecules is given by the following expression:

$$\text{Interaction energy, } \phi(r) = -\frac{\mu_1^2\alpha_2}{(4\pi\varepsilon_0)^2 r^6} \quad ...(2)$$

where μ_1 refers to the permanent dipole moment of molecule 1, α_2 refers the polarizability of molecule 2 and r, as usual, refers to the separation of the molecules. $4\pi\varepsilon_0$ as mentioned above, refers to the permittivity factor.

3. Induced Dipole—Induced Dipole Interaction. London Forces or Dispersive Forces

The van der Waals forces are also known to exist even in non-polar molecules like O_2 ad N_2 and also in non-polar monoatomic molecules like He, Ne, Ar, tc. This attraction has been evident from the condensation of these gases into liquids at sufficiently high pressures and low temperatures. This could not be understood for may years.

In 1930, F. London gave a satisfactory explanation for the existence of forces of attraction between *no-polar* molecules. This explanation is based on quantum mechanics. According to this view, electrons of a neutral molecule keep on oscillating with respect to the nuclei of the atoms. Because of this, at a given instant, positive charge will be concentrated in one region and negative charge in another region of the same molecule. Hence, a non-polar molecule would become momentarily self polarised. This polarised molecule maybe able to induce a dipole moment in a neighbouring molecule with antiparallel orientation, as depicted in Fig. 4.19.

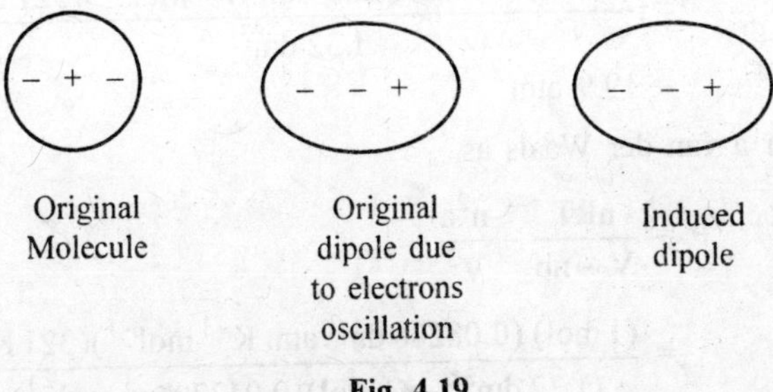

Fig. 4.19

The electrostatic forces of attraction between induced dipoles and the original dipoles (due to electron oscillation) are termed as London forces. These forces are also termed as dispersive forces because the well known phenomenon of dispersion of light is also related to these dipoles.

For a pair of adjacent molecules, London forces have been found to vary inversely as the *seventh power* of the distance between them, *i.e.*, $F \propto 1/r^7$. The magnitude of these forces, evidently, would be very appreciably small.

The vån der Waals attraction is *non-polar molecules is therefore exclusively ascribed* to London forces. In polar molecules having permanent dipoles, however, other electrostatic forces because of dipole-dipole interaction also contribute towards the van der Waals attraction.

SOLVED PROBLEMS

Example 1:

Calculate the pressure exerted by one mole of carbon dioxide gas in a 1.32 dm³ vessel at 48°C using :

(a) the ideal gas equation and

(b) the van der Waals equation. The van der Waals constants are;

$$a = 3.59 \ dm^6 \ atm \ mol^{-2}$$

and $$b = 0.0427 \ dm^3 \ mol^{-1}.$$

Solution:

(a) For an ideal gas, PV = nRT, so that

$$P = \frac{nRT}{V}$$

$$= \frac{(1\ \text{mol})\ (0.08206\ \text{dm}^3\ \text{atm K}^{-1}\ \text{mol}^{-1})(321\ \text{K})}{1.32\ \text{dm}^3}$$

$= 19.9$ atm.

(b) For a van der Waals as,

$$P = \frac{nRT}{V - nb} - \frac{n^2 a}{V^2}$$

$$= \frac{(1\ \text{mol})\ (0.08206\ \text{dm}^3\ \text{atm K}^{-1}\ \text{mol}^{-1})(321\ \text{K})}{(1.32\ \text{dm}^3) - (1\ \text{mol})(0.0427\ \text{dm}^3\ \text{mol}^{-1})}$$

$$- \frac{(1\ \text{mol})^2(3.59\ \text{dm}^6\ \text{atm mol}^{-2})}{(1.32\ \text{dm}^3)^2}$$

$= 2062$ atm $- 2.06$ atm $= 18.56$ atm

The experimental value of the pressure is 16.40 atm. Thus, van der Waals equation gives a better agreement with experiment than the ideal gas equation.

Example 2:

Two moles of ammonia are enclosed in a five litre flask at 27°C. Calculate the pressure of ammonia using van der Waals equation. For ammonia, a = 4.17 atm. litre^{-2} mole^{-2} and b = 0.0371 litre mole^{-1}.

Solution:

V is the volume of n (= 2) moles of the gas = 5 litres.

a = 4.17 atms litre^{-2} mole^{-2} (correction due to inter-molecular attraction).

b = 0.0371 litre mole^{-1} (volume correction)

R = 0.082 litre atms. absolute degree^{-1} mole^{-1}.

T = 27 + 273 = 300MK

number of moles (n) = 2

According to van der Waals equation,

$$\left(P + \frac{a}{V^2}\right)(V - b) = Rt, \text{ valid for one mole.}$$

$$\left(P + a\frac{n^2}{V^2}\right)(V - nb) = nRT \text{ valid for n moles.}$$

i.e.
$$P = \frac{nRT}{V - nb} - \frac{an^2}{V^2}$$

Substituting the values of a, b, R, V and T in equation.

$$P = \frac{2 \times 0.082 \times 300}{5 - 2 \times 0.0371} - \frac{4.17 \times 2 \times 2}{5 \times 5}$$

$$= 9.98 - 0.67 = 9.31 \text{ atm.}$$

Example 3:

A vessel of 50 litre capacity contains 10 moles of steam under 5 atmospheres of pressure. Calculate the temperature of the steam using van der Waals equation if for water a = 5.46 atm (litre/mole)² and b = 0.031 litre/mole.

Solution:

We know $\left(P + \frac{n^2a}{V^2}\right)(V - nb) = nRT$

i.e.
$$T = \left(P + \frac{n^2a}{V^2}\right)\left(\frac{V - nb}{nR}\right)$$

Here $P = 5$ atm, $n = 10$ moles

$V = 10$ litres and $a = 5.46 \text{ atm} \left(\frac{\text{litre}}{\text{mole}}\right)^2$

and $b = 0.03$/lite/mole.

Substituting the values,

$$T = \left(5 + \frac{(10)^2 \times 5.46}{(50)^2}\right)\left(\frac{50 - 10 \times 0.031}{10 \times 0.08}\right)$$

$$= 311.9°K$$

Example 4:

The critical pressure and critical temperature of a gas obeying van der Waals equation have values 73 atms and 37°C. Calculate constants 'a' and 'b'. The value of the gas constant, R is 0.082 atm degree⁻¹.

Solution:

We have $$T_c = \frac{8a}{27Rb} \quad ...(i)$$

$$P_c = \frac{a}{27b^2} \quad ...(ii)$$

Dividing (i) by (ii) we have,

$$\frac{T_c}{P_c} = \frac{8a}{27Rb} \times \frac{27b^2}{a} = \frac{8b}{R}$$

Now $T_c = 31° + 273 = 304°K$

$P_c = 0.082$ atm degree^{-1}

Hence $$\frac{304}{73} = \frac{8b}{0.082}$$

$\therefore$ $b = 0.4269$

Substituting the value of b in (i) *i.e.*,

$$T_c = \frac{8a}{27Rb}, \text{ we get}$$

$$304 = \frac{8a}{27 \times 0.082 \times 0.04269}$$

$a = 3.591.$

Example 5:

Calculate the reduced pressure, volume and temperature of CH_4 gas when one mole of gas is contained in a 5 litre flask under 5 atmospheric pressure. What will be the temperature of the as when

Critical temperature $T_c = 190.2°K$

Critical volume $V_c = 0.0988$ *litre/mole and*

Critical pressure $P_c = 45.6$ *atm.*

Solution:

We know,

Reduced pressure $\pi = P/Pc$

Here P = 5 atm $P_c = 45.6$ atm.

Substituting the values, $\pi = 5/45.6 = 0.1096$

Reduced volume $\phi = V/V_c$

Here V = 5 litres and V_c = 0.0988 litre/mole

substituting the values $\phi = \frac{5}{0.0988} = 50.607$

According to the law of corresponding states,

$$\left(\pi + \frac{3}{\phi^2}\right) \times (3\phi - I) = 8\theta$$

where θ is the reduced temperature

i.e. $$\theta = \frac{\left(\pi + \frac{3}{\phi^2}\right) \times (3\phi - I)}{8}$$

Substituting the values of π and φ,

$$\theta = \frac{\left[0.1096 + \frac{3}{(50.607)^2}\right](3 \times 50.607 - 1)}{8}$$

$$= 2.0888$$

We know that reduced temperature

$$\theta = \frac{T}{T_c}$$

$$T = T_c \times \theta$$

Here $T_c = 190.2°K$

Substituting the values,

Temperature = 190.2 × 2.0888 = 397.3°K.

Some Examples are given in Four Types

Type I: *Calculation of velocity when temperature alone is given.*

By kinetic gas equation:

$$PV = \frac{1}{3} mn\bar{c}^2$$

$$(R.M.S.V.) = \sqrt{\frac{3PV}{mn}} = \sqrt{\frac{3PV}{M}} = \sqrt{\frac{3RT}{M}}$$

M denotes the molecular weight of the gas and R = 8.3×10^7 ergs

$$(\text{RMSV}) = \sqrt{\frac{3 \times 8.3 \times 10^7 \times T}{M}}$$

$$= 1.58 \times \sqrt{\frac{T}{M}} \times 10^4 \text{ cm/sec.}$$

Example 6:

For a fluid show that (¶P/¶T)$_V$ = a/b where a is expansivity and b is compressibility of the gas. Hence, show that for an ideal gas.

$$(\partial P/\partial T)_V = R/V$$

Solution:

As a gas can be defined by any two of the three parameters P, V and T, we can write

$$V = f(T\,.\,P)$$

As V is a state function, hence its total differential can be written as follows:

$$dV = \left(\frac{\partial V}{\partial T}\right)_P dT + \left(\frac{\partial V}{\partial P}\right)_T dP \quad ...(1)$$

By definition, $\alpha = \frac{1}{V}\left(\frac{\partial V}{\partial T}\right)_P$ so that $\left(\frac{\partial V}{\partial T}\right)_P = \alpha V$

and $\beta = -\frac{1}{V}\left(\frac{\partial V}{\partial P}\right)_T$ so that $\left(\frac{\partial V}{\partial P}\right) = -\beta V$

On substituting for α and β in Eq. (i) we get

$$dV = \alpha\, VdT - \beta\, VdP$$

When V is constant, dV = 0. Hence,

$$\alpha\, VdT - \beta\, VdP = 0$$

or $V(\alpha\, dT - \beta\, dP) = 0$

But $V \neq 0$, hence

$$\alpha\, dT - \beta\, dP = 0$$

or $\alpha\, dT = \beta\, dP$

or $\left(\frac{\partial P}{\partial T}\right)_V = \frac{\alpha}{\beta}$

For an ideal gas we know, $\alpha = 1/T$ and $\beta = 1/P$. Hence,

$$\left(\frac{\partial P}{\partial T}\right)_V = \frac{\alpha}{\beta} = \frac{1/T}{1/P} = \frac{P}{T} = \frac{R}{V}$$

This is to be proved.

Example 7:

Calculate the root mean square and average velocity of carbon dioxide at 1000°C.

Solution:

$$(\text{RMSV}) = 1.58 \times \sqrt{\frac{T}{M}} \times 10^4 \text{ cm/sec.}$$

Now T = 273 + 1000 = 1273 K; M = 44

$$\therefore \quad (\text{RMSV}) = 1.58 \times \sqrt{\frac{1273}{44}} \times 10^4 \text{ cm/sec.}$$

$$= 84870 \text{ cm/sec.}$$

This gives the root mean square velocity.

Now average velocity = 0.9213 × R.M.S.V. of CO_2

Average velocity = 0.9213 × 84870 = 78190 cm/sec.

Type II : *Calculation of molecular velocity at N.T.P.*

From the kinetic gas equation we have derived that

$$\text{RMSV} = \sqrt{\frac{3PV}{M}}$$

Knowing the pressure and volume at N.T.P. we can calculate the value of RMSV as exemplified below.

Example 8:

Calculate the root mean square velocity of oxygen at N.T.P.

Solution:

For oxygen at N.T.P.

P = 76 cms = 76 × 13.6 × 981 dynes/sq-cm

V = 22.4 litres = 22400 cc.

Substituting the values in the equation:

$$(\text{RMSV}) = \sqrt{\frac{3PV}{M}}$$

we have, $$RMSV = \sqrt{\frac{3 \times 76 \times 13.6 \times 981 \times 22400}{32}}$$

$$= 46140 \text{ cm/sec.}$$

Type III : *Calculation of molecular velocity when pressure and temperature. (i.e., not at N.T.P.) are given.*

In such calculations, the first step involves the conversion of gram molecular volume (22.4 litres at N.T.P.) to the volume under the different pressure and temperature. Having found the new volume, this value is substituted in the equation:

$$RMSV = \sqrt{\frac{3PV}{M}}$$

This calculation will become clear from the following example:

Example 9:

Calculate the root mean square velocity of nitrogen at 27°C and 700 mm pressure.

Solution:

(i) Conversion of G.M.V. of new volume

At N.T.P.	At New Conditions
$P_1 = 760$ mm	$P_2 = 700$ mm
$V_1 = 22400$ c.c.	$V_2 = ?$
$T_1 = 273$ °K	$T_2 = 273 + 27 = 300$ °K

By gas equation

$$\frac{P_1V_1}{T_1} = \frac{P_2V_2}{T_2}$$

$$V_2 = \frac{P_1V_1}{T_1} \times \frac{T_2}{P_2}$$

$$= \frac{22400 \times 760}{273} \times \frac{300}{700} = 26720 \text{ c.c.}$$

(ii) Calculating the value of RMSV.

Now $$P = 700 \text{ mm} = 70 \text{ cm}$$

$$= 70 \times 13.6 \times 981 \text{ dynes/cm}$$

$$V = 26720 \text{ c.c.}; M = 28 \text{ gm}$$

Substituting the values in the equation:

$$\text{RMSV} = \sqrt{\frac{3PV}{M}}$$

$$\text{RMSV} = \sqrt{\frac{3 \times 70 \times 13.6 \times 981 \times 26720}{28}}$$

$$= 51710 \text{ c/sec.}$$

Type IV : *Calculation of molecular velocity when pressure and density are given.*

$$PV = \frac{1}{3}mnc^2 = \frac{1}{3}Mc^2$$

$$\text{RMSV} = \sqrt{\frac{3PV}{M}} = \sqrt{\frac{3P}{D}}$$

where $\frac{V}{M}$ denotes density (D) of the gas in gm/c.c.

Example 10:

Oxygen has a density of 1.429 gm. per litre at N.T.P. Calculate the R.M.S. and A.V. of oxygen molecules.

Solution:

Now $\quad \text{RMSV} = \sqrt{\frac{3P}{D}}$

In this case, $\quad P = 1$ atmosphere

$$= 76 \times 13.6 \times 981 \text{ dynes/sq. cm.}$$

$$D = 1.422 \text{ gm/litre}$$

$$= \frac{1.429}{1000} \text{ gm/c.c} = 0.001429 \text{ gm/c.c.}$$

Substituting the values in the above equation, we have

$$\text{RMSV} = \sqrt{\frac{3 \times 76 \times 13.6 \times 981}{0.001429}} = 46100 \text{ cm/sec}$$

This is the R.M.S. (root mean square mean velocity) of oxygen molecules.

Now A.V. (Average velocity)

$$= 0.9213 \times \text{R.M.S. velocity}$$

$$= 0.9213 \times 46{,}100 = 42458 \text{ cm/sec.}$$